# Analysis, design and construction of steel space frames

*Dedication*

Professor G. S. Ramaswamy dedicates this book to his wife Seetha, his constant companion, counsellor and critic.

M. Eekhout dedicates this book to the pioneers Z. S. Makowski, G. S. Ramaswamy, K. Venkateswarlu, N. Hedge and H. Nooshin who introduced space frame construction into India and promoted its applications during the last decade.

G. R. Suresh dedicates this book to his wife Sangeeta and his parents.

# Analysis, design and construction of steel space frames


G. S. Ramaswamy

*Civil Engineering Consultancy Consortium, Chennai, India*

M. Eekhout

*Technological University, Delft, and Octatube International BV, Delft, The Netherlands*

G. R. Suresh

*EDS, Sydney, Australia*

Co-authors:

M. Papadrakakis and N. D. Lagaros (Chapter 9)
*Institute of Structural Analysis and Seismic Research, National Technical University, Athens, Greece*

S. Rajasekaran (Chapter 10)
*PSG College of Technology, Coimbatore, India*

F. G. A. Albermani (Chapter 10)
*University of Queensland, St. Lucia, Queensland, Australia*



Thomas Telford

Published by Thomas Telford Publishing, Thomas Telford Ltd, 1 Heron Quay, London E14 4JD
URL: http://www.thomastelford.com

Distributors for Thomas Telford books are
*USA*: ASCE Press, 1801 Alexander Bell Drive, Reston, VA 20191-4400
*Japan*: Maruzen Co. Ltd, Book Department, 3–10 Nihonbashi 2-chome, Chuo-ku, Tokyo 103
*Australia*: DA Books and Journals, 648 Whitehorse Road, Mitcham 3132, Victoria

First published 2002

A catalogue record for this book is available from the British Library
ISBN: 0 7277 3014 2



Typeset by Helius, Brighton and Rochester
Printed and bound in Great Britain by MPG Books, Bodmin

# Contents

# Foreword

The space frame is one of the most important structural systems very often adopted in modern buildings of large dimensions. Among other structural systems the space frame has a few essential advantages. One of those advantages is its lightness. The lightness of the structure is due mainly to the fact that material is distributed spatially in such a way that the load transfer mechanism is primarily axial, either tension or compression, so that in any given element all material is fully utilized. In large-span roofs, where the self-weight of the structure constitutes an important part of the total load, the lightness of the constituent elements largely contributes to the rationality and economy of the whole building.

Another advantage of the space frame is that the production and construction techniques are industrialized to an extent much greater than is the case in conventional structural systems. The linear elements and the joints are both prefabricated, so that the jointing work at the site is relatively simple. Thus a space frame can sometimes be erected by semi-skilled workers. The light weight of the individual elements also makes the handling and assembly work easier.

A space frame is sufficiently stiff in spite of its lightness. This is due to its three-dimensional character and to the full participation of the constituent elements, which adapt themselves equally well to almost all types of loading.

This book deals with the whole process of the realization, namely the analysis, design and construction, of a space frame. It is rare that a book on a structural system develops its coverage to such an extent as in this book. Those who begin reading this book with only a fundamental knowledge of structures will, before they finish reading the major part of it, find themselves provided with sufficient knowledge for further study of space frames to become experts in research, design and realization.

This book is authored by extraordinarily qualified experts of space structures. Professor G. S. Ramaswamy is one of the most well-known researchers of spatial structures in and outside India. He has been very active in the International Association for Shell and Spatial Structures for a long time, currently being a member of its Advisory Board. Professor M. Eekhout is also a very famous expert of spatial structures in The Netherlands. His books *Tubular Structures in Architecture* and *Architecture in Space Structures* are well known worldwide. G. R. Suresh is a talented and well-trained postgraduate structural engineer currently working with EDS in Sydney.

I hope that this book will be read by many structural engineers who would like to contribute to the development of space frames through their work in analysis, design and construction.

*Professor Mamoru Kawaguchi, Dr.Eng., Dr.-Ing.E.h.*
*Hosei University, and President of the International Association for Shell & Spatial Structures*

# Preface

This book is an outgrowth of and a spin-off from the consulting activities of Professor G. S. Ramaswamy and Professor Dr Ir M. Eekhout, who have had the oppurtunity of designing and building steel space frames for diverse applications in Europe, the Middle East, India, the Far East and Australia.

The genesis of the book may be traced to 1989 when Professor Ramaswamy organized an *ad hoc* international consortium to design and build the steel space frame roof over the sports hall of the Al Wahda Club at Abu Dhabi. Professor Ramaswamy, while shopping around for a suitable node connector for this project, as the consultant to Nagarjuna Coated Tubes Ltd (NCTL) of Hyderabad who had bid for the job, was put in touch with Mick Eekhout by their mutual friend Professor Makowski, who is generally acknowledged as the foremost promoter of modern steel space frames. Professor Ramaswamy had known and worked with Makowski since his days at the Battersea Polytechnic. He was the PhD examiner of Professor Eekhout, who regards him as one of his mentors. For the Abu Dhabi project, Professor Ramaswamy assumed the responsibility for the conceptual and preliminary design. He also co-ordinated the inputs from the different partners. The final designs and working drawings were done at Professor Eekhout's office at Delft. The proprietary Tuball node connectors were supplied by Octatube Space Strcutures BV, The Netherlands, which is headed by him. The tubes were fabricated at NCTL, Hyderabad, and shipped to Dubai. The assembly and erection at Abu Dhabi were carried out by Galadari Engineering works of Dubai under the general supervision of the Consolidated Construction Company. A collaboration of this kind with the partners based in Delft, Dubai and India would have been unthinkable had not the fax machine appeared on the scene at about that time. Professor Eekhout's abiding interest in Indian art and architecture and Professor Ramaswamy's exposure to other cultures during his stint as a United Nations Advisor in Iraq, Saudi Arabia and the West Indies and as a professor in the USA were also factors that contributed to the success of this unique international experiment.

Based on the experience gained on this job, it was concluded that costs could be drastically cut if the node connectors were manufactured in India to stringent Dutch specifications and the designs were done at Professor Ramaswamy's office at Chennai and checked at Delft. Fortunately, India has several ISO 9002 approved foundries of world class, and there is no dearth of highly skilled computer-oriented design engineers. This judicious strategy of combining proven European technology with lower Indian manufacturing and design costs did pay rich dividends, and in the following years the Consortium won several contracts in the United Arab Emirates for projects involving space frames, against stiff international competetion. Some of these projects are presented in this book in the form of case studies. While the case studies documented here had necessarily to be tied to the Octatube system, other well-known competing systems are also described in the book, with illustrative examples of their application.

Although steel space frames originated in Europe, of late the centre of gravity of applications has moved Eastward to the Middle East, India, China and Japan. This movement from the Western developers to Eastern markets has taken place at a time when space frames are an establised product in Europe and are no longer seen by Western architects as state of the art. However, in developing countries, with a time setback of about 30 years, they are still viewed as a novelty. Moreover, the kind of applications being attempted in developing countries are somewhat different from the trends in Europe. In the West, space frames were generally adopted for one of their types of application. But in India and China, with ambitious plans of industrialization on the anvil, involving large numbers of similar sheds, steel space frames are often standardized and the factory-made components are shipped to remote locations for assembly and erection by relatively unskilled labour. There is no compromise on quality insofar as the components are concerned, because they are manufactured in factories with stringent quality control. The only deviation from European practice is that simple dericks and pulley blocks are often used to hoist the components at remote locations where cranes are not always easily available. In countries such as the United Arab Emirates, where almost everything is imported, space frame construction proceeds in exactly the same manner as in Western Europe.

Professor Ramaswamy is indebted to Professor Nooshin for introducing him to the elegance and power of Formex formulation and the use of the allied Formian software. Chapter 5 bears the imprint of the interaction he had with Professor Nooshin during his consultancy missions to India. His invitation to Professor Ramaswamy to act as the Guest Editor of the special issue on aircraft hangars of the *International Journal of Space Structures* fortuitously brought him into contact with Professor Manolis Papadrakakis and Dr Nikolaos D. Lagaros, who co-authored Chapter 9 of this book.

Professor Rajasekaran, who had often collaborated with Professor Ramaswamy in the past, readily accepted his invitation to co-author Chapter 10, Stability Checks. The authors are indebted to him for his lucid exposition of a complex subject. An invitation to visit the University of Queenland, Brisbane, Australia, to act as an external examiner for a PhD thesis provided an opportunity for Professor Ramaswamy to collaborate with Dr Albermani, and this led to his involvement in the non-linear analysis of the Pragati Maidan Dome as a co-author of Chapter 10.

This book is believed to be the first of its kind to present in one volume the analysis, design and construction aspects of all types of space frame. Addressed to consultants, senior undergraduate and post graduate students, and the research community, the presentation is tailored to suit their varying levels of prior knowledge of the subject. The book does not assume any knowledge of mathematics and mechanics beyond what is taught in a first degree course in engineering. All formulae are derived from first principles.

A unique feature of the book is the fully documented case studies that accompany the chapters. Fully worked examples and some useful computer programs, provided on the CD-ROM that accompanies this book, may be easily retrieved and viewed with the help of user-friendly GUI features.

Professor Ramaswamy records his gratitude to Dr Narayanan, Director of the Structural Engineering Research Centre, Chennai, for his invaluable assistance in putting the manuscript together. The authors thank Messers Surya Prakash, T. Anand, P. B. Karthick, D. Chander Dev and C. S. Venkata Subramanian for their assistance with the preparation of the text, drawings and the CD-ROM.

The authors express their grateful appreciation to MERO, Raumstruktur GmbH & Co., Würzburg, Germany, and CORUS, UK, for making photographs available and granting us permission to quote from their publications. Their thanks are also due to many authors, publishers and firms for granting permission to use material from their publications.

*G. S. Ramaswamy*
*M. Eekhout*

# About the CD-ROM

## Contents

The CD-ROM contains:

- SAP90 input and output data for the forces, reactions and displacements in some of the case studies
- the executable programs mentioned in the text, with illustrative examples
- figures relating to some of the case studies and a graphical comparison of the formulae prescribed in the British, Indian, Australian, European and American Codes for the design of compression members.

## Retrieval of contents

User-friendly GUI features facilitate easy access to the CD-ROM and retrieval of the stored information. The GUI is written in HTML and is designed to work with Microsoft Internet Explorer (version 4.0 or later), which must be installed on your computer to use this CD-ROM. (Note: Internet Explorer is not provided on this CD-ROM but can be downloaded from Microsoft's Web site (www.microsoft.com).)

## Procedure

- Insert the CD-ROM in the drive.
- Open Internet Explorer.
- From the File menu, choose Open and click Browse.
- From the dialog box, choose the CD-ROM drive from the 'Look in' dialogue box.
- From the list of files appearing in the dialogue box, select Index.htm and then click Open.
- The selected file will appear in the 'Open' dialogue box. Now, click OK.
- The GUI will now lead you through the rest of the process.

# 1. Introduction to space frames

## 1.1. Introduction

Space structures are a leaf taken from Nature. Natural forms possess exceptional rigidity and use minimum materials to maximum structural advantage. To quote Makowski, 'Natural forms act in the direction of minimum effort'. In 1940, Le Recolais, a well-known exponent of space frames, drew attention to the geodesic type of spherical surfaces that characterize the bone skeletons of *Radia tuscaretta globosa*. Man has not been slow in copying these examples drawn from Nature.

It may not perhaps be well known that Alexander Graham Bell, the inventor of the telephone, was one of the first to appreciate the merits of space frames and apply them in actual construction. In fact, he built one of the early versions of a flying machine using a multi-layered space frame concept. As the saving in weight is more critical in aircraft structures than in buildings, it is not surprising that the earliest efforts to take advantage of the lightness of space frames, stemming from their structural efficiency, were in building flying machines. It is in this context that the Wright brothers discovered the possibilities offered by the tetrahedron.

Historically, the earliest space structures to be built were the domes pioneered by the German Engineers Schwedler and Zimmermann in the nineteenth century. Zimmermann achieved a span of 361 ft for an Exhibition Structure built at Lyons as early as 1894. The Schwedler-type dome built over the Civic Centre at Charlotte, North Carolina, USA, immediately after the Second World War had a clear span of 332 ft. These indeed were remarkable feats for those times. These early experiments were all but forgotten until 1937, when Dr Mengeringhausen revived interest in space frames. He realized that the promotion of space frames on a large scale would be possible only if the system lent itself to factory production and field assembly. To achieve these twin objectives, it was necessary to reduce variety and develop a means of easy assembly. Ideally, a space frame must consist of members of identical length, although a practical compromise is often to settle for as few different lengths as possible. Dr Mengeringhausen's pioneering topological studies were aimed at reducing variety and his invention of the MERO connector elegantly solved the problem of easy assembly. With these developments, Mengeringhausen was able to successfully commercialize space frames in 1942. His universal MERO node connector (Fig. 1.1) can accept as many as 18 members at different angles in tapped

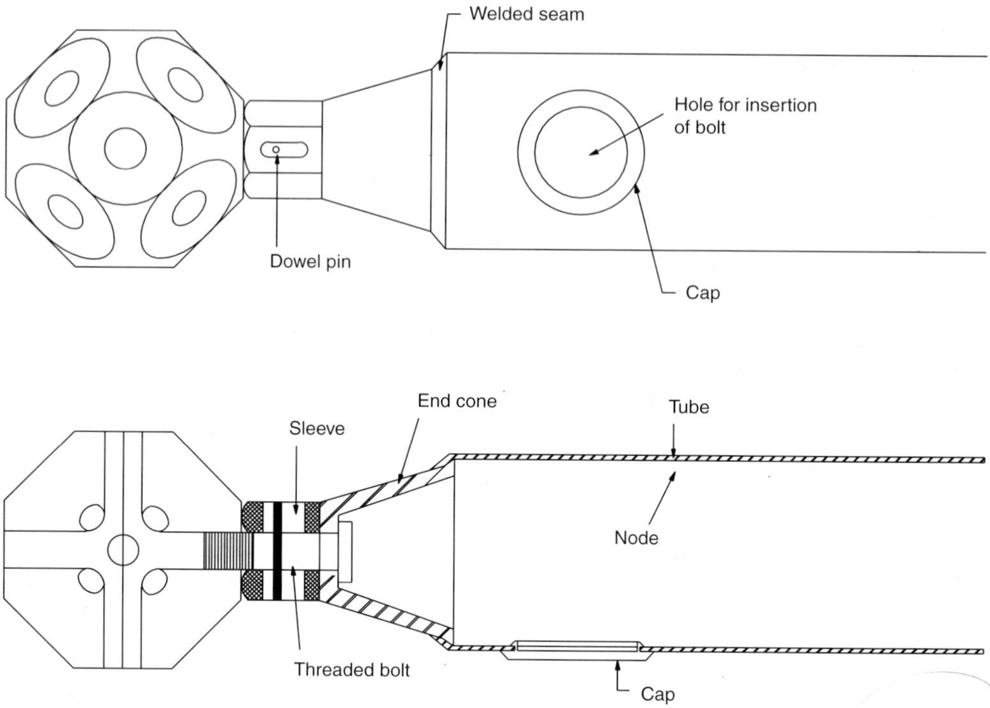


*Fig. 1.1. The MERO system. (Reproduced with permission from Hiroyuki [1.13])*

holes distributed over its surface. It became possible, for the first time, to build space frames out of precise mass-produced factory-made components that could be assembled in simple Mechano fashion. Transportation posed no problems, because the individual components were of manageable size and weight. Complex space frames can be built even at remote locations using relatively unskilled labour. Mengeringhausen's work triggered similar developments all over the world. The following years witnessed a proliferation of proprietary systems. Octatube, Nodus, Unistrut, Unibat, Space Deck and Triodetic and their variations are among the systems that were developed during this period in different parts of the world. Some of these systems are described in later paragraphs in this chapter.

## 1.2. Definition of a space frame

Drawing on the report on the state-of-the-art issued in 1984 by the International Association for Shell and Spatial Structures (IASS) [1.1], one may adopt the following definition:

> A Space Frame is a structural system, assembled of linear elements so arranged that the loads are transferred in a three-dimensional manner. In some cases, the constituent elements may be two-dimensional. Macroscopically, a space frame often takes the form of a flat or curved surface.

In the 1976 American Society of Civil Engineers report entitled *Lattice Structures: State-of-the-Art Report* [1.2] space frames are referred to as lattice structures and are defined as:

> ... a structural system in the form of a network of elements (as opposed to a continuous surface) ... another characteristic of lattice structural systems is that their load-carrying mechanism is three dimensional in nature.

A distinction is sometimes drawn between space frames and space trusses. According to this terminology, *space trusses* are those systems that are pin-jointed, while the term *space frame* is reserved for structures with rigid joints. In this chapter, following the IASS Working Group's report, 'space frame' is used as a generic term, of which space trusses are merely a subset.

## 1.3. Advantages of space frames

- They are light, structurally efficient and use materials optimally.
- They are an elegant and economical means of covering large column-free spaces for a variety of applications, including sports structures, auditoria, aircraft hangars and assembly halls.
- They carry loads by three-dimensional action. Loads applied at a point are not merely carried by the members meeting at that joint but are dispersed to be shared by a large number of other members. It is thus possible to transfer heavy concentrated loads to any joint. This characteristic is especially useful in aircraft maintenance hangars, because it permits docking systems to be hung from nodes, providing unlimited flexibility.
- Deflections are small because of the high inherent stiffness. This characteristic is taken advantage of in using space frames as back-up structures for parabolic dish antennas [1.3], multi-plate antennas [1.4] and telescopes [1.5], which demand high rigidity as well as lightness. The accuracy required for such applications is so high that the tolerances are specified in terms of the root mean square of the deviations from the ideal surface. Space frames meet these exacting requirements.
- Because space frames are put together by using precise, factory-made components, unskilled labour is adequate for their assembly and erection.
- Services, such as lighting and air conditioning, can be integrated with space frames and they can often be installed on the ground itself, thus obviating the hazards of working at heights.
- Space frames save construction time, because they use factory-produced components that can be manufactured by fast production techniques, transported to the site and easily erected.
- Each of the components is light, facilitating transportation.
- They offer the architect unrestricted freedom in locating supports and planning the subdivision of the covered space.
- Being light structures, dead loads are very much less and there are consequent savings in columns and substructures.

## 1.4. Shells and space frames

Space frames are akin to shells in structural behaviour, indeed so much so that they may be described as 'sisters under the skin'. These similarities may be summarized as follows:

- Loads applied are primarily carried by axial forces (tension or compression) with little or no bending.
- Deflections are small because of the inherent stiffness.
- Both are efficient structural forms, optimizing the use of materials.

Because of these close similarities in structural behaviour, preliminary analysis of space frames can often be carried out by the continuum analogy, involving the replacement of a space frame by an equivalent shell or plate.

## 1.5. Components of a space frame

A space frame consists of *axial members*, which are preferably tubes, also known as circular hollow sections or rectangular hollow sections, and *connectors*, which join the members together. Other structural sections such as I- and H-sections are also occasionally used, especially if loads that are transferred to the members between nodes cause bending in addition to axial forces. Where loads are applied only at nodes, circular and rectangular hollow sections have an edge over other section types because they are more efficient in compression, offering a higher radius of gyration for the same area. Circular hollow sections have the further advantage that their moments of inertia are the same in all directions.

## 1.6. Members

The members of a space frame are either of aluminium or steel. Aluminium members are of light weight but are more expensive. The large majority of space frames use members of steel with a yield strength ranging from 210 to 450 MPa. The tubes may be electrically resistance welded or seamless.

The Allied & Conduit Corporation, USA, manufactures a proprietary electrical resistance welded steel tube known as a Flocoat tube, with a yield strength of 450 MPa. The unique process involves the use of steel strips cold-rolled into a precision tube which is continuously hot-dip galvanized followed by the on-line application of a protective coat to make it rust-proof. This process is shown schematically in Fig. 1.2.

The protective coats applied on-line over the galvanizing consist of a chromate conversion coat followed by a clear polymer coating. This proprietary tube has an edge over ordinary steel tubes in two respects:

- it has very high corrosion resistance
- it has a higher yield strength of 450 MPa.

When Flocoat tubes are not available, it is recommended that steel tubes and connectors be galvanized and powder-coated to the colour of the client's choice if superior corrosion resistance and aesthetics are desired.

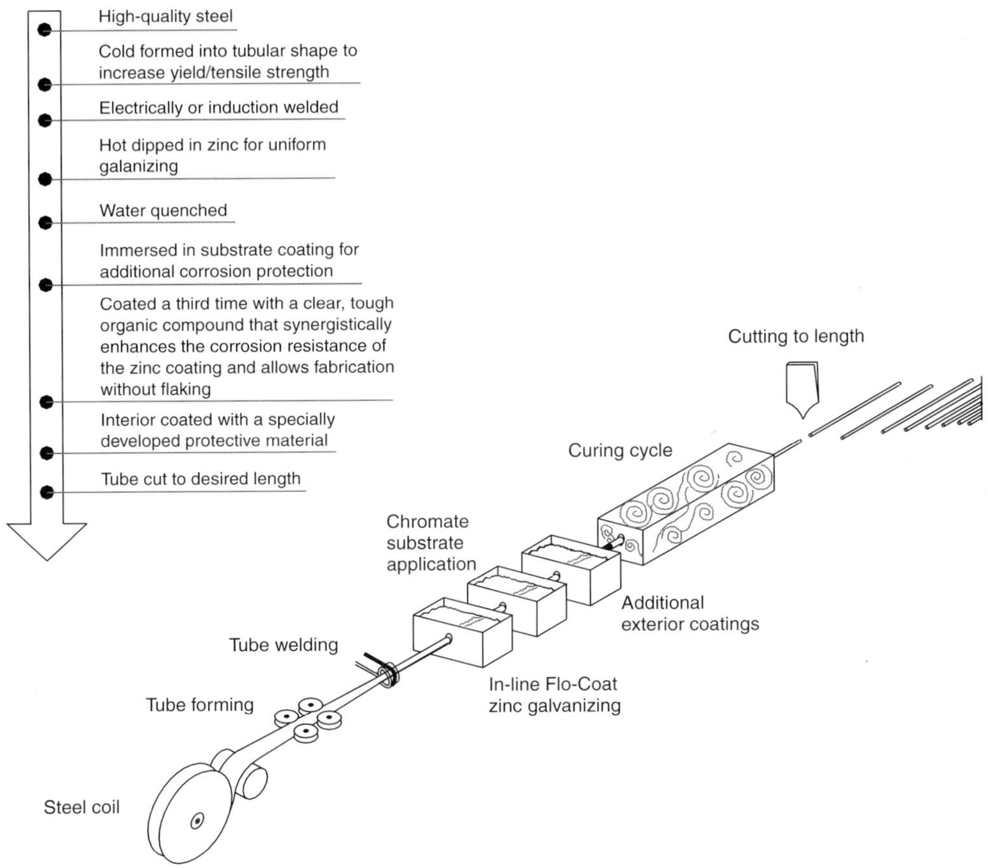


*Fig. 1.2. The Flocoat process*

When on-line galvanizing is not possible during manufacture, it is the normal practice to subject tubes to hot-dip galvanizing, both inside and out, to ensure adequate corrosion resistance. The tubes may be powder-coated to provide additional protection.

## 1.7. Node connectors

MERO is an abbreviation for Mengeringhausen, the inventor of the connector. The MERO connector (see Fig. 1.1) consists of the following components:

- A threaded spherical ball of hot forged steel with as many as 18 tapped holes, at different angles, distributed evenly over its surface, to receive tubular members at different angles. The sphere has flat surfaces around the threaded holes to improve the seating of the spanner sleeve. The holes are precisely drilled so that the centre lines of the tubes at a node meet at the centre of the sphere.
- A bolt, which is inserted through a hole in the tubular member and passes through a cone welded to the end of the tube.

- A hexagonal spanner sleeve.
- A dowel pin, which goes through the threaded bolt and connects it to the spanner sleeve.

The following sequence is involved in installing the tube in the node:

- The tubes are cut to the correct lengths.
- The end cones are welded to both ends of the tube.
- The member with the end cones is galvanized.
- The bolts are inserted through the holes at the end of the tubes.
- The inserted bolts are connected to the spanner sleeves by the dowel pin. The spanner sleeve is also galvanized.
- The bolts are driven into the drilled holes in the spherical ball using the spanner sleeve. To improve the appearance, the holes in the tubes provided for inserting the bolts may be covered with a plastic cap. The connector shown in Fig. 1.1 is suitable only for transmitting axial loads, and it is primarily used for double-layer grids.

## 1.8. New generation of MERO node connectors

In recent years MERO has introduced four new node connectors that are particularly suitable for single-layer, shell-type space frames, which need flexural rigidity at the nodes to improve stability. These connectors are:

- the cylindrical node connector ZK
- the plate-disc node connector TK
- the hollow hemi-spherical node connector NK
- the block node connector BK.

For more information on the MERO system, the MERO catalogue [1.6] may be consulted.

## 1.9. The Octatube and Tuball node connectors

The Octatube node connector developed by Prof. Dr Ir. Mick Eekhout of The Netherlands consists of an octagonal base plate to which are welded two semi-octagonal plates placed at right angles to each other. The Octatube node connector is a plate connector (Fig. 1.3). The tubes meeting at a node are flattened and connected by means of high-strength bolts. Developed in 1973, this node connector can be manufactured in any well-equipped workshop. The connector is designed for space frames meant to roof workshops, warehouses and other structures where cost rather than aesthetics is the governing consideration. It is possible to use sections other than tubes to effect the connection if a plate is welded to the end of the member. The space frame roof of the arrival hall at Bangalore Airport was built using this system (Fig. 1.4).

The Tuball connector, developed by Eekhout in 1984 [1.7], is a hollow sphere made of spheroidal graphite. One-fourth of the sphere comprises a cap and the rest is a cup (Fig. 1.5). The end of the circular or rectangular hollow section member to

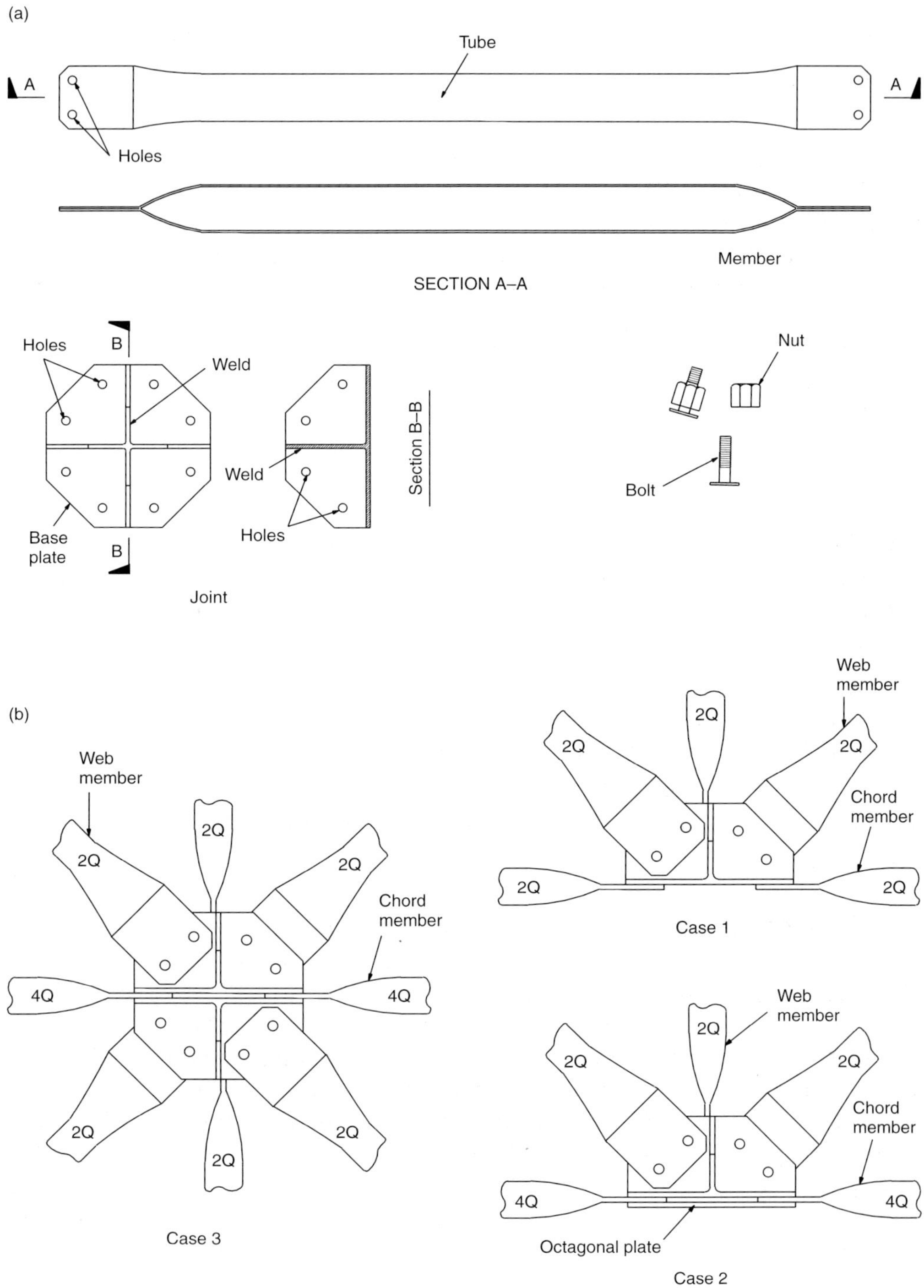


*Fig. 1.3. The Octatube system: (a) the components; (b) assembly patterns of the nodes. (Reproduced with permission from Hiroyuki [1.13])*

be connected is fitted at its ends with threaded solid props by welding. Working from inside the cup, high-strength bolts, normally of 8.8 or 10.9 grade, according to the British Standard Code, are driven into the threaded prop by means of a torque wrench. Coning of the ends of the tube is resorted to if tubes of large diameter are to be accommodated without congestion (Fig. 1.6) over the surface of the cup. If the tension to be transmitted to a node exceeds the permissible tensile strength of the node, the member is carried through the node by using a threaded rod to connect the ends of the tubes (see Fig. 1.6). Being hollow, the Tuball node tends to be lighter than a solid forged node. It is also less expensive, because spheroidal graphite costs less than forged steel. The Tubal node has been used successfully for building numerous space frames in India, the United Arab Emirates and the Far East using relatively unskilled labour.

## 1.10. The Nodus system

The Nodus connector (Fig. 1.7) is a precision engineered product developed by the Tubes Division of the British Steel Corporation [1.8]. Its advantages are that it can accept both rectangular and circular hollow sections and that the cladding can be fixed directly to the chords. This is possible because the node is flush with the chords. The two half-casings, which are the principal components of the system, are made of spheroidal graphite of grade SNG 27/12 conforming to BS 2789 with a Charpy impact value at 0°C of not less than 4.5 ft-lb and a 0.5% proof stress of not less than 19 t/m$^2$. The chord connectors have to be welded to the ends of the hollow members on site using a specially designed jig.

*Fig. 1.4. The space frame over the arrival hall at Bangalore Airport, Bangalore, India*

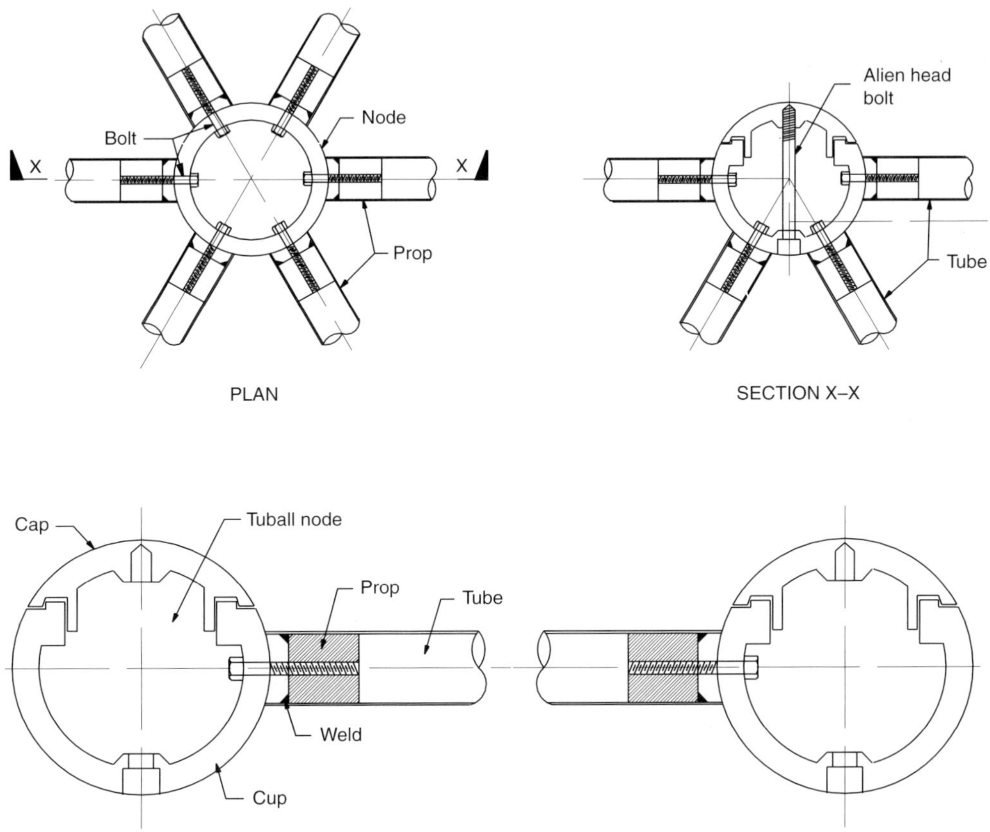


*Fig. 1.5. The Tuball connector*

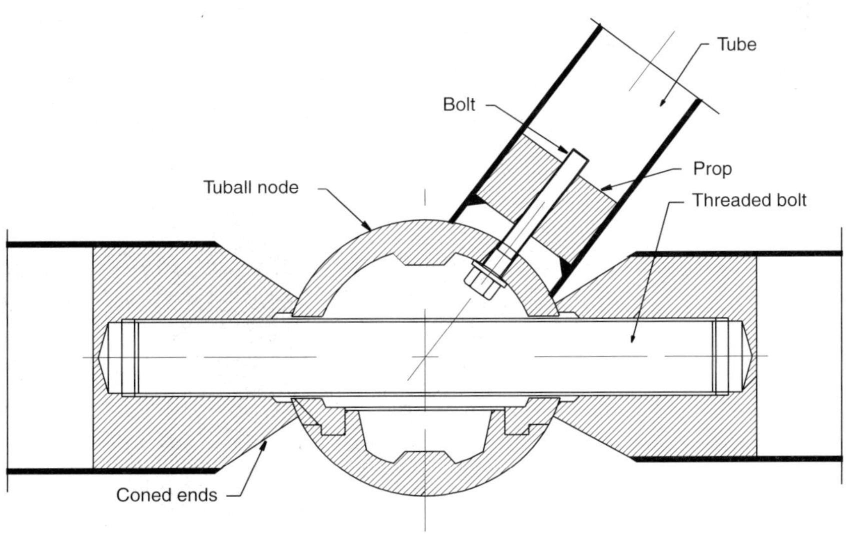


*Fig. 1.6. A Tuball node with a threaded rod and coned props*

The space frame over the Gatwick railway station concourse in England is a good example of the application of the Nodus system (Fig. 1.8). The design brief called for a 63 m by 33 m concourse with no interior columns. The pre-set parameters and the need to integrate the concourse with the grid and orientation of the British Airport Authority's building at the concourse level limited the options open to the architect and structural engineer. Causing minimum inconvenience to passengers and to British Rail was a major consideration that led to the choice of the Nodus space frame. Assembling the space frame on the ground and lifting it into position was not feasible at Gatwick due to the restricted space available. The structure was therefore assembled *in situ* on scaffolding.

This multi-layer space frame utilized 84 t of structural hollow sections (88.9 mm × 3.2 mm and 114.3 mm × 6.3 mm circular hollow sections; 120 mm × 120 mm × 8.0 mm rectangular hollow sections), mostly of grade 43C with a small amount of grade 50C steel. In total 725 nodes were used. The engineer for the project was Mr F. S. Proctor, Chief Engineer, British Rail, Southern Region, the main contractor was Cementation Construction Ltd, and the steelwork subcontractor was Fairfield Malery Ltd. More detailed information on the project is available elsewhere [1.9].

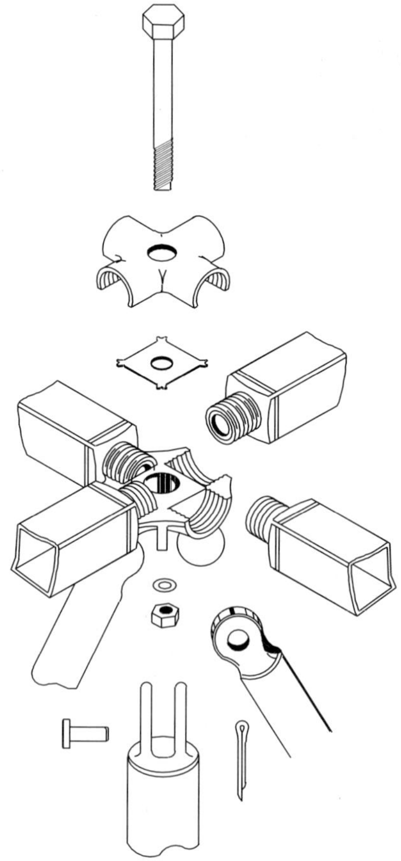

*Fig. 1.7. The Nodus joint. (Reproduced with permission from Hiroyuki [1.12])*

*Fig. 1.8. The Nodus space frame over the concourse of the Gatwick railway station*

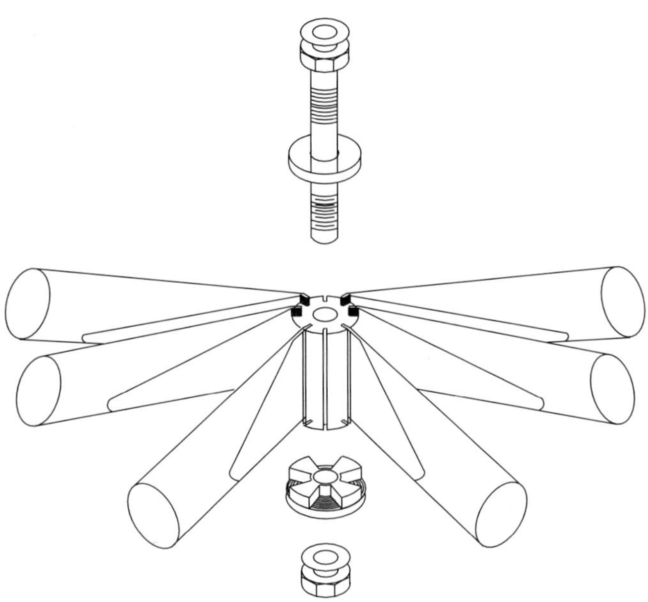

*Fig. 1.9. The Triodetic joint. (Reproduced with permission from Hiroyuki [1.13])*

## 1.11. The Triodetic connector

The Triodetic connector, due to A. E. Fentiman, is a Canadian development. It consists of a hub, usually an aluminium extrusion, that has slots or key ways, and the ends of members are pressed or coined to match the slots (Fig. 1.9). The member is

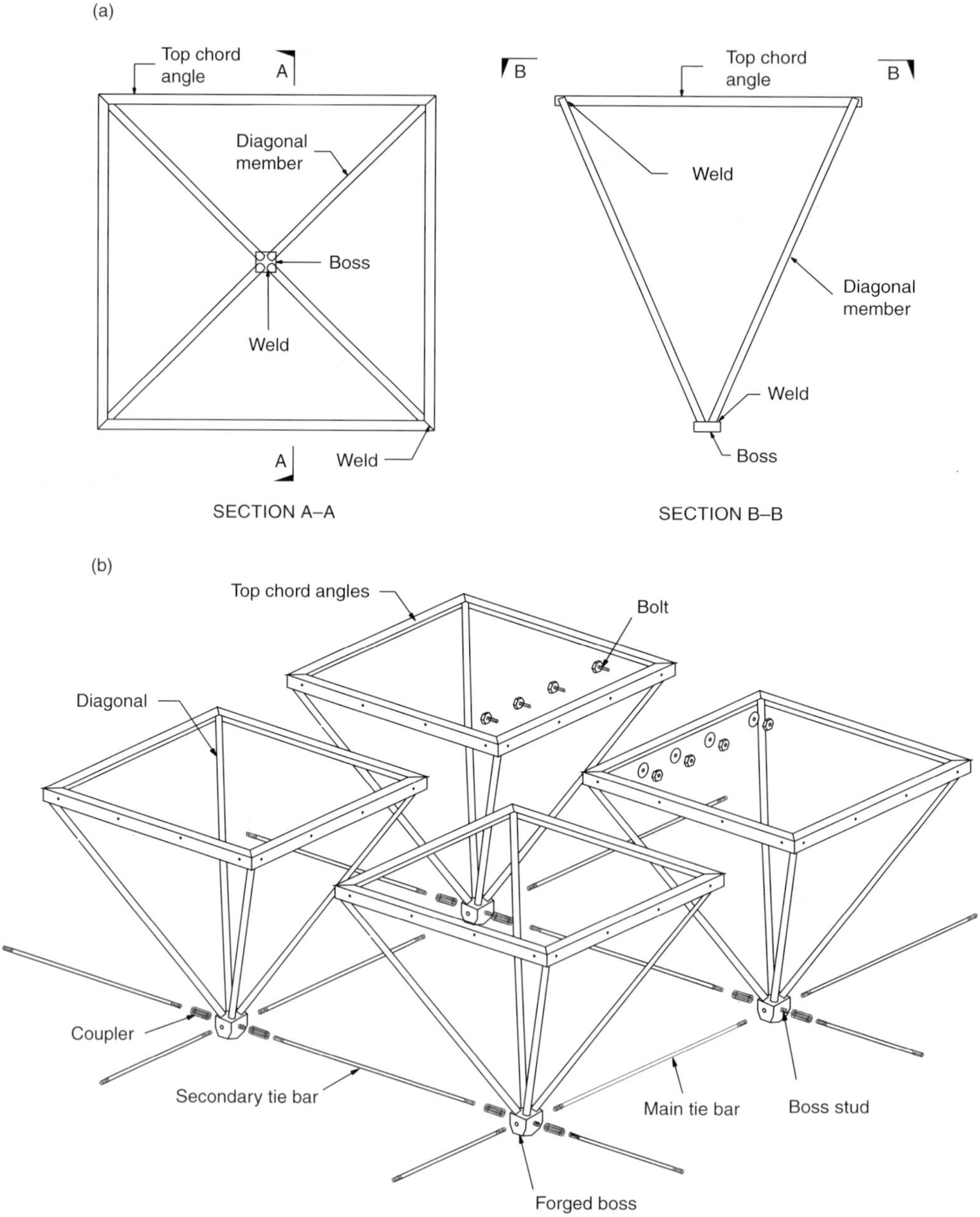


*Fig. 1.10. The Space Deck unit: (a) the components; (b) the standard assembly. (Reproduced with permission from Hiroyuki [1.13])*

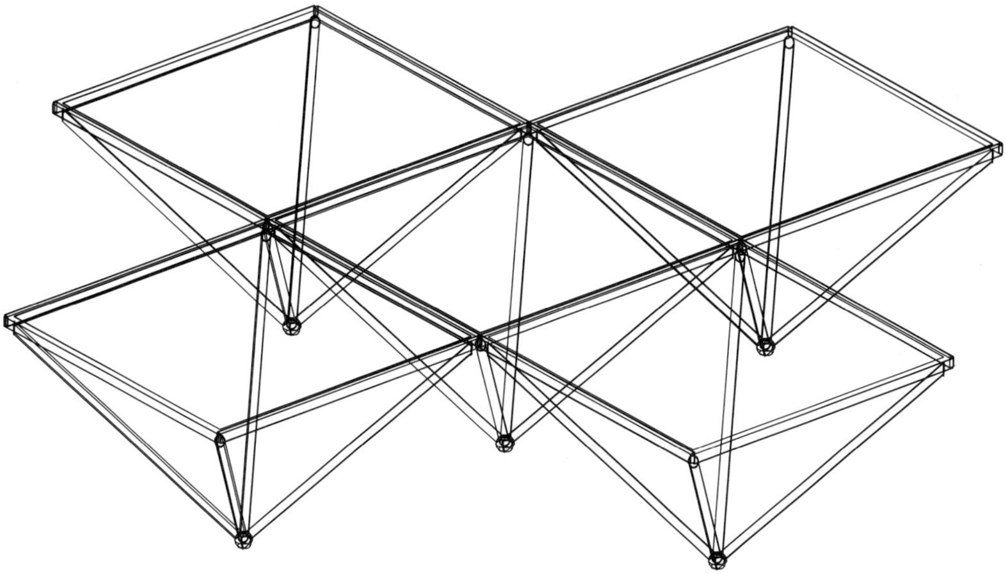

*Fig. 1.11. The Unibat system. (Reproduced with permission from Hiroyuki [1.13])*

inserted into the slot with slight pressure. The central stud and retaining washer hold the assembly together. A hub can accommodate 12 members. It is claimed that the connector can be used with steel as well as aluminium tubes. It is stated that if steel tubes conform to SAE 1010 and SAE 1020 (US designation) or 65ST6 (Canadian designation), there is no risk of electrolytic corrosion [1.10]. Most of the structures built using this system consist of aluminium tubes and aluminium connectors [1.11].

## 1.12. Modular systems

Space frame systems may be roughly categorized as nodular and modular systems. The systems described so far in this chapter are *nodular systems*, which consist of members and nodes. In *modular systems*, instead of members there are prefabricated units, normally in the form of pyramids, which are joined together by means of high-strength friction bolts. Two well-known modular systems are described below.

### *1.12.1. The Space Deck system*

The components of the Space Deck system are shown in Fig. 1.10. Developed by the British firm Space Deck, the system has been in use since 1954. It consists of factory-produced inverted pyramids composed of the top chords, and four diagonal bracing members. The pyramids are connected together by means of bottom tie members and bolts.

### *1.12.2. The Unibat system*

The components of the Unibat system and the manner in which they are put together are shown in Fig. 1.11. The four top chord members and the four diagonal

*Fig. 1.12. The platform shelter at Thirumailai Station, LUZ, Chennai, India*

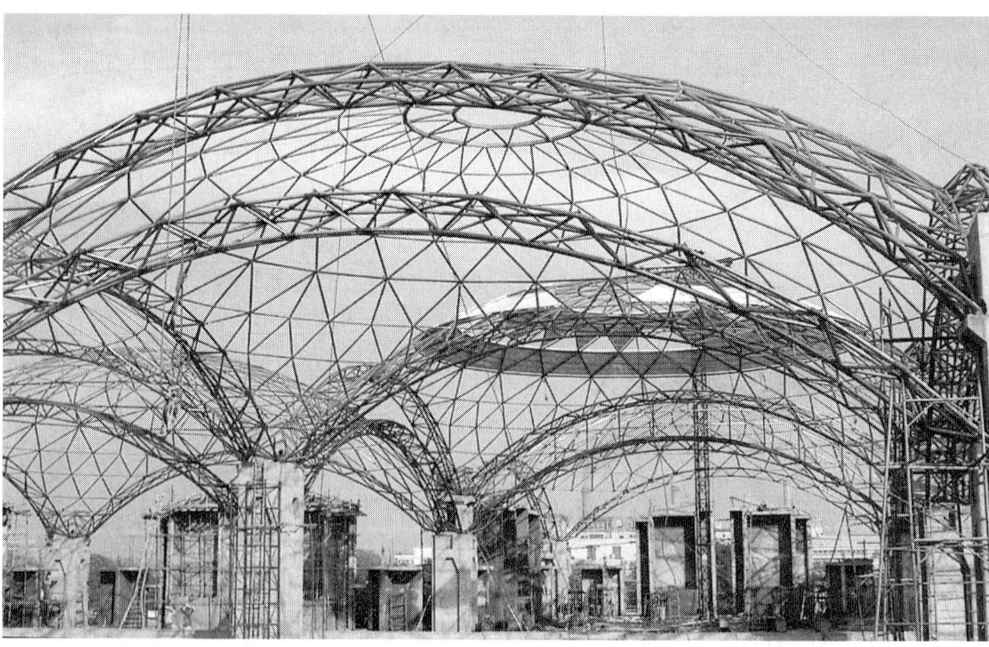

*Fig. 1.13. The dome over the exhibition hall at Pragati Maidan, for the Trade Fair Authority, New Delhi, India*

*Fig. 1.14. The space frame over the domestic cooking gas bottling plant, for the Indian Oil Corporation, Cuddapah, India*

*Fig. 1.15. The Haj Terminal, Jeddah, Saudi Arabia*

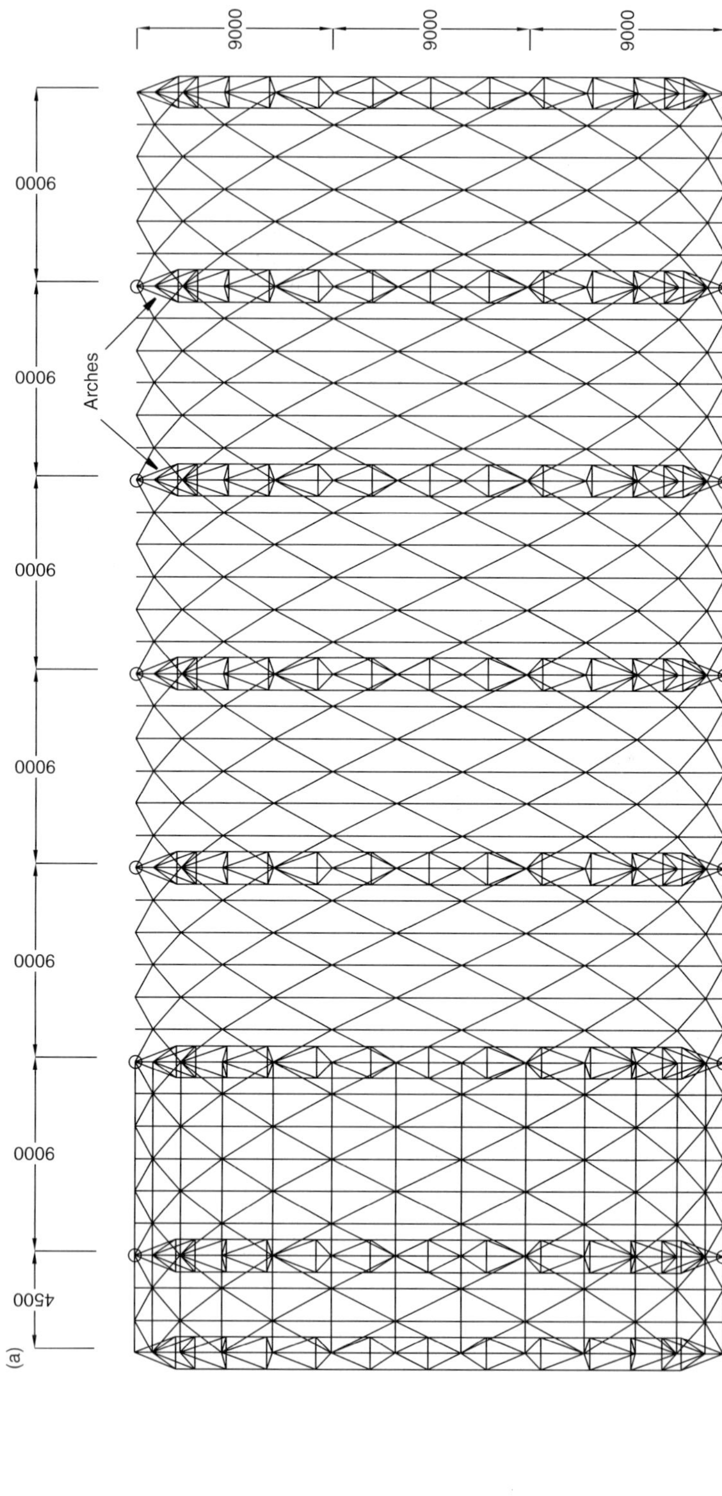
9000
9000
9000
9000
9000
9000
9000
9000
9000
4500
Arches
(a)

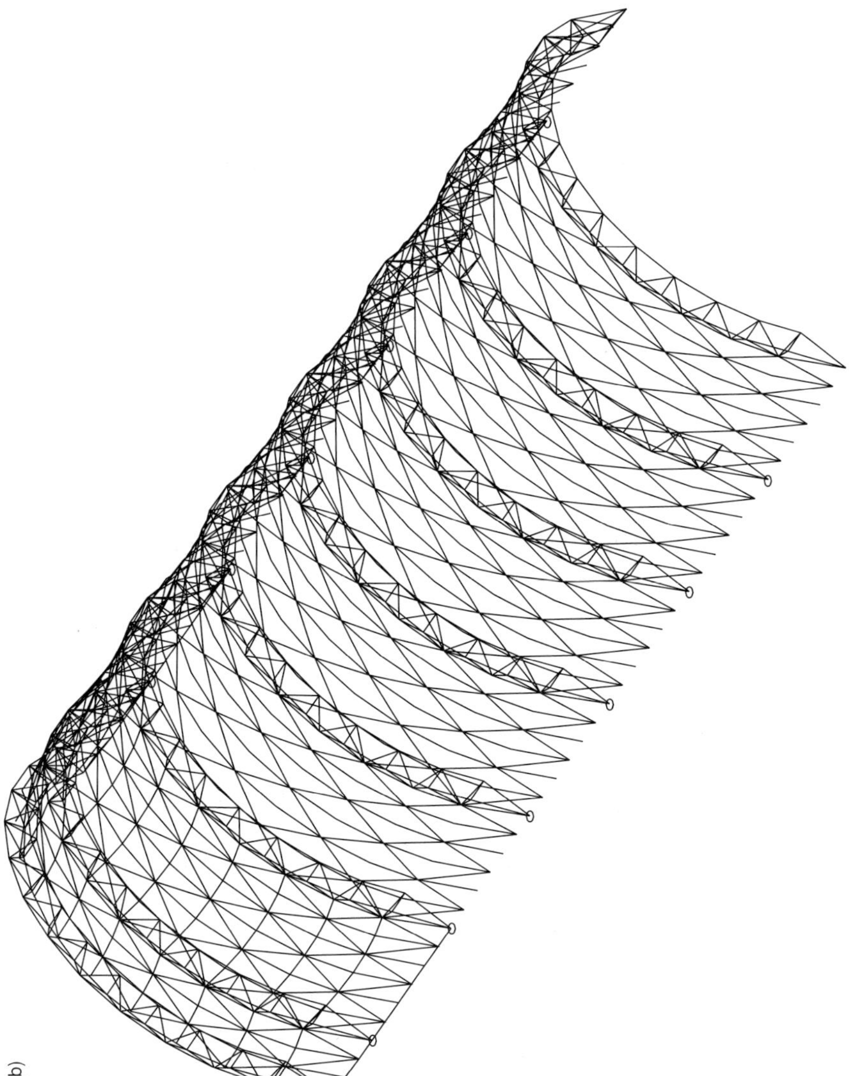


*Fig. 1.16. The roof of the Maritime Museum built for the port of Singapore Authority by L&M Prestressing PTE Ltd: (a) the frame plan; (b) cross-section*

bracings comprise the factory-made inverted pyramid. Forged pieces at the four corners and an apex piece at the bottom permit the modules to be connected together by means of high-strength friction bolts. Modular units are quicker to assemble in the field and they eliminate expensive node connectors.

## 1.13. New node connectors

A number of new space frame systems have recently become available on the Japanese market. Among these are the Tomoe Unit Truss, the TM Truss, the NS Space Truss, the SS Space Truss and the KT Truss. A detailed description of these is given in a paper by Kawaguchi [1.12]. A comprehensive survey of nodular and modular systems has recently been presented by Hiroyuki Tomatsuri [1.13].

## 1.14. Types of space frame

### *1.14.1. Two- and three-way grids*

Space frames are characterized as two-way or three-way, depending on whether the members intersecting at a node run in two or three directions.

### *1.14.2. Single-, double- and triple-layered space frames*

A single-layer space frame has to be singly or doubly curved. Barrel vaults (Fig. 1.12) and domes (Fig. 1.13) are examples. By far the most common space frames are double layered and flat (Fig. 1.14). The structural depth of a space frame increases with its span. For very large spans the structural depth becomes large, as does the length of the diagonal bracing members. In such instances a triple-layer space frame may be an appropriate solution. The middle layer, being located at the neutral surface, has little or no structural function, but it does reduce the length of the bracing members, making it possible to use lighter sections. Multiple-layer, flat space frames are discussed in detail in Chapter 6. A triple-layer space frame with a five-layer front girder of 129 m unsupported span used to roof the recently completed maintenance hangar for Air India at Mumbai is described in Chapter 6 .

## 1.15. Fabric structures

The development of modern fabric structures may be traced to the pioneering investigations of Frei Otto at the Institutus für leichte Flächen Tragwerke, University of Stuttgart [1.14]. The fabric normally used is woven from fibreglass yarn coated with PTFE (polytetrafluoroethene), commonly known as Teflon). This material was developed by the CHEMFAB Corporation, USA, in association with Du Pont Company and Owens Corning Fibreglass. The first fabric roof using this new material was built for the student centre at La Verne College, California, USA, in 1972. The PTFE used to coat the fabric is one of the most chemically inert of plastics. Its low heat transmission characteristics, long service life (estimated at 25 years), translucency, and ability to withstand ultraviolet radiation, moisture and pollutants makes Teflon-coated fibreglass an ideal roofing material. In wide-span

*Fig. 1.17. The fabric structure of the Maritime Museum, Singapore*

roofing applications the fabric may be reinforced using cables and supported on poles or suspended from overhead cables.

The most outstanding application of cable-reinforced fabric structures suspended from overhead cables is the Haj Terminal at the Abdul Aziz International Airport at Jeddah, which was built to accommodate the unusually high traffic during the Haj Pilgrimage [1.15]. The terminal covers 105 acres, has a covered area of 500 000 $m^2$ and can accommodate 50 000 pilgrims. Designed by Skidmore, Owings and Merril, the terminal comprises 210 tent-shaped roofs, each 45 m by 45 m in plan, formed from suspended, cable-reinforced, Teflon-coated, fibreglass fabric (Fig. 1.15). The modules, which are catenoids on square ground plans, reflect the vernacular tent architectural tradition. The translucent, heat-insulating fabric lets in diffused light and provides a comfortable ambience in the desert. However, Teflon-coated fibreglass fabric is expensive, and PVC-coated polyester, which is self-extinguishing, meets all the code requirements and is often preferred as a cheaper alternative.

Cable-reinforced fabric structures similar to those used to roof the Haj Terminal involve non-developable surfaces that pose several complex problems. These relate to the development of fabric cutting patterns and sealing of the seams. These problems may be overcome by draping the pretensioned fabric over a developable cylindrical surface such as a barrel vault, as was done for two structures built for the Port of Singapore Authority to house a maritime museum and an exhibition hall. These structures involve two braced cylindrical barrel vaults of size 27 m × 57.8 m and 31.2 m × 83.8 m. The structural arrangement adopted was the same for both vaults. The barrel vaults involve double-layered arches spaced 9 m apart (Fig. 1.16). The structural designs of L&M pre-stressing were proof-checked by the first author for Permafab Pty Ltd, Australia, who supplied the Teflon-coated, glass-fibre fabric for the project. A view of the finished museum is shown in Fig. 1.17.

The imaginative combination of space frames, cable-stays and fabrics opens up an unlimited array of possibilities for developing elegant and cost-effective solutions for enclosing large column-free spaces for a diversity of applications.

## 1.16. References


1.1. Analysis, design and realization of space frames: a state-of-the-art report. *Bulletin of the International Association of Shell and Spatial Structures, Special Issue*, 1984, **15**(84/85).

1.2. TASK COMMITTEE ON LATTICED STRUCTURES, COMMITTEE ON SPECIAL STRUCTURES, COMMITTEE ON METALS, STRUCTURAL DIVISION. Latticed structures: state-of-the-art report. *Journal of the Structural Division, ASCE*, 1976, **102**(ST11).

1.3. RAJARAMAN, A., APPA RAO, T. V. S. R, KHADAKKAR, A. G. and RAMASWAMY, G. S. Minimum weight designs of parabolic antennas. *Proceedings of the 2nd International Conference on Space Structures*, University of Surrey, Guildford, 1975, pp. 804–809.

1.4. LACOPULOS, A. J. Space frames for multiple antenna reflecting panels. *Proceedings of the International Conference on Space Structures*, University of Surrey, Guildford, 1966, pp. 1134–1143.

1.5. MEDWADOWSKI, S. J. Space frames in architecture and science. *Bulletin of the International Association for Shells and Space Structures*, 1996, **XXVII-3**(92/93); 1997, **XXVIII-1**, 7–12.

1.6. EEKHOUT, M. *Architecture in Space Structures*, UITGEVERIJ 010, Delft, 1989, p. 85.

1.7. MERO – Raumstruktur GmbH Co., Würzburg, Postfach 6169, D-8700 Würzburg, Germany.

1.8. BRITISH STEEL CORPORATION. *NODUS Space Frame Grids. Part 3: Construction*, Structural Advisory Service, Tubes Division, Croydon, 1984.

1.9. BRITISH STEEL CORPORATION. *Tubular Structures 34*, Structural Advisory Service, Tubes Division, Corby.

1.10. FENTIMAN, H. G. Development in Canada in the fabrication and construction of three-dimensional structures using the triodetic system. *Proceedings of the First Conference on Space Structures*, University of Surrey, Guildford, 1966, pp. 1073–1082.

1.11. ELLIOT, A. W. Triodetic domes. In: Z. S. Makowski (ed.) *Analysis, Design and Construction of Braced Domes*, Granada, London, 1984, ch. 25.

1.12. KAWAGUCHI, M. A review of the recent development of steel space structures in Japan. *Bulletin of the International Association for Shell and Spatial Structures*, 1988, **XXIX-1**(96).

1.13. TOMATSURI HIROYUKI. Space structure forms and systems, M.Phil. thesis, University of Surrey, Guildford, 1994, ch. 4.

1.14. INSTITUTUS FÜR LEICHTE FLÄCHEN TRAGWERKE. *IL 16: Zelte* [Tents], University of Stuttgart, 1976.

1.15. GOLDSMITH, M. Effects of scale on long span roofs. *Proceedings of the 3rd International Symposium on Wide Span Structures*, University of Stuttgart, 1985, vol. 1, pp. 1–4.

# 2. Structural design of space frame components

## 2.1. Introduction

In the design of members, the *limit state approach* has been adopted. Loads corresponding to the ultimate limit state are applied at the nodes and a linear elastic analysis is carried out. The examples that follow have been worked out in accordance with the provisions of the Eurocode 3 recommendations [2.1] or BS 5950: Part 1: 1985 [2.2], both of which are written in the limit state format. These codes find wide acceptance in the UK, Europe, the Middle East and the Far East.

## 2.2. Structural sections used in space frames

One may use tubes, also known as circular hollow sections or rectangular hollow sections. The tubes may be electrical resistance welded or seamless.

## 2.3. Design of tension members

According to BS 5950: Part 1, the area of cross-section required is found by dividing the tension in the member at the ultimate limit state by the yield strength of the steel. If, for example, the tension in a space frame member at the ultimate limit state is 90 kN and if the yield stress is 450 MPa, the area of cross-section required is

$$A = 90 \times 1000/450 = 200\ \text{mm}^2 \tag{2.1}$$

## 2.4. Design of compression members

The design of compression members according to Eurocode 3 may be carried out by using the following step by step procedure.

*Step 1*

From the yield stress of the steel $\rho_y$ and its Young's modulus $E$, compute the non-dimensional parameter

$$\lambda_e = \pi\sqrt{\frac{\rho_y}{E}} \tag{2.2}$$

If $E = 210$ kN/mm$^2$ and $\rho_y = 360$,

$$\lambda_e = \pi\sqrt{\frac{210\,000}{360}} = 75.90 \tag{2.3}$$

*Step 2*

Compute the slenderness ratio $\lambda_{rel}$ of the member from the following data, which refer to a space frame member:

yield strength = 360 MPa
outside diameter of tube = 76.2 mm
thickness = 3.65 mm
actual length, $l = 3529.00$ mm
effective length, $l_e = 0.85 \times 3529 = 3000.00$ mm
radius of gyration, $r = 25.65$
area = 832.00 mm$^2$
$l_e/r = \lambda = 3000/25.65 = 116.9$ mm
$\lambda_{rel} = \lambda/\lambda_e = 1.54$

*Step 3*

Eurocode 3 gives two different formulae for arriving at $\omega_{buc}$, the coefficient by which the yield strength is to be multiplied to arrive at the permissible stress in compression at the ultimate limit state. From these formulae $\omega_{buc}$ can be computed and tabulated. The formulae incorporate factors to account for initial imperfections. Values of $\omega_{buc}$ are given in Table 2.1. It is conservative practice to use the values for cold-rolled sections for all tubes.

Using the 'Program for Calculating the Allowable Compression and Tension at the Ultimate Limit State for Cold Rolled Sections – Eurocode B' availabe on the CD-ROM, we arrive at an allowable compression of 97.9498 kN.

*Step 4*

The allowable compression in the tube at the ultimate limit state is

$$\rho_y A \omega_{buc} = 360 \times 832 \times 0.3277 = 98 \text{ kN} \tag{2.4}$$

## 2.5. Effective length of compression members

The effective length $l_e$ of a compression member may be found by multiplying its actual length $l$ by a coefficient $K$:

$$l_e = Kl \tag{2.5}$$

The value of $K$ may be assumed to be 1.0 for members with plate connectors and 0.85 for members with spherical connectors. Alternatively, the effective length may be taken as the actual length of the tube between the faces of the two adjacent connectors.

*Table 2.1. Values of* $\omega_{buc}$

| $\lambda_{rel}$ | $\omega_{buc}$ Hot-rolled sections (Eurocode 3, Table a) | $\omega_{buc}$ Cold-rolled sections (Eurocode 3, Table b) |
|---|---|---|
| 0.2 | 1.0000 | 1.0000 |
| 0.3 | 0.9775 | 0.9641 |
| 0.4 | 0.9528 | 0.9261 |
| 0.5 | 0.9243 | 0.8842 |
| 0.6 | 0.8900 | 0.8371 |
| 0.7 | 0.8477 | 0.7837 |
| 0.8 | 0.7957 | 0.7245 |
| 0.9 | 0.7339 | 0.6612 |
| 1.0 | 0.6656 | 0.5970 |
| 1.1 | 0.5960 | 0.5352 |
| 1.2 | 0.5300 | 0.4781 |
| 1.3 | 0.4703 | 0.4269 |
| 1.4 | 0.4179 | 0.3817 |
| 1.5 | 0.3724 | 0.3422 |
| 1.6 | 0.3332 | 0.3079 |
| 1.7 | 0.2994 | 0.2781 |
| 1.8 | 0.2702 | 0.2521 |
| 1.9 | 0.2449 | 0.2294 |
| 2.0 | 0.2229 | 0.2095 |
| 2.1 | 0.2036 | 0.1920 |
| 2.2 | 0.1867 | 0.1765 |
| 2.3 | 0.1717 | 0.1628 |
| 2.4 | 0.1585 | 0.1506 |
| 2.5 | 0.1467 | 0.1397 |
| 2.6 | 0.1362 | 0.1299 |
| 2.7 | 0.1267 | 0.1211 |
| 2.8 | 0.1182 | 0.1132 |
| 2.9 | 0.1105 | 0.1060 |
| 3.0 | 0.1036 | 0.0994 |

## 2.6. Design of members under axial compression and bending

Ideally, purlins should transfer loads to space frames only at nodes. Occasionally, it may happen that the grid module, representing the distance between adjacent nodes, is too large for the roofing sheets to span, necessitating the use of intermediate purlins to rest directly on the member. If this is the case, the member has to be designed to carry a bending moment in addition to an axial force. We consider here the case where the axial force is a compression $N$. Consider the beam-column shown in Fig. 2.1 subjected to the action of an axial compression $N$ and a lateral load $q(x)$, causing a bending moment $M_1(x)$ at a section a distance $x$ from the origin. The corresponding deflection $w_1(x)$ may be expanded as a Fourier series. Restricting ourselves to only the first term,

$$w_1(x) = w_1 \sin\frac{\pi x}{l} \tag{2.6}$$

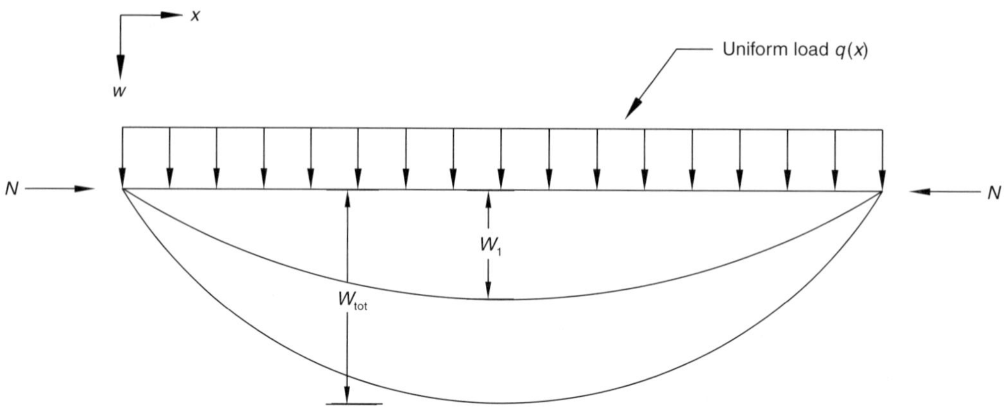


*Fig. 2.1. A beam column*

$$M_1(x) = EI\omega_1 \frac{\pi^2}{l^2} \sin\frac{\pi x}{l} \tag{2.7}$$

where $E$ is Young's modulus and $I$ is the moment of inertia of the member.

The maximum value of $M_1(x)$, occurring at the centre of the span, is

$$\frac{\pi^2 EI}{l^2} w_1 \tag{2.8}$$

The axial load will cause an additional bending moment, which will increase the central deflection to $w_{tot}$.

Hence, the total bending moment at the centre is $M_1(x) + Nw_{tot}$.

We may represent the total deflection at $x$ as

$$w_{tot} = w_{tot} \sin\frac{\pi x}{l} \tag{2.9}$$

Hence, the bending moment at the centre is given by

$$\frac{\pi^2 EI}{l^2} w_1 + (N) w_{tot} = w_{tot}\left(\frac{\pi^2 EI}{l^2}\right) \tag{2.10}$$

Let $F_u = \pi^2 EI/l^2$. Then, from equation (2.10),

$$w_{tot} = \frac{F_u w_1}{F_u - N} \tag{2.11}$$

Letting $F_u/N = n$,

$$w_{tot} = \left(\frac{n}{n-1}\right) w_1 \tag{2.12}$$

It follows that

$$M_{tot}(x) = \left(\frac{n}{n-1}\right) M_1(x) \tag{2.13}$$

The parameter $n$ may be regarded as a bending moment magnification factor. An example would best illustrate the use of the equation.

*Example 2.1*

Select the size of a tubular member of a space frame to carry an axial compression of 30 kN and a lateral central concentrated load of 2 kN at the ultimate limit state. The length of the member is 2800 mm and the yield strength is 210 MPa.

If $E = 210\,000$ MPa and $K = 0.85$, and using Eurocode 3, Table b (see Table 2.1), try a tube of size 88.9 mm × 4.05 mm. The tube properties are then:

Area, $A = \pi(88.9 - 4.05) \times 4.05 = 1079 \text{ mm}^2$
Radius of gyration, $r = (88.90 - 4.05)/2\sqrt{2} = 29.90$ mm
Moment of inertia, $Ar^2 = 964\,636 \text{ mm}^4$
Modulus of section $= (964\,636 \times 2)/88.90 = 21\,701 \text{ mm}^2$
$F_u = \pi^2 EI/l^2 = (\pi^2 \times 210 \times 964\,636)/2800^2 = 255$ kN
$n = F_u/N = 255/30 = 8.50$
$n/(n-1) = 8.5/7.5 = 1.13$
$M_{to} = (2000 \times 2800 \times 1.13)/4 = 1\,582\,000$ N-mm
Bending stress $= 1\,582\,000/21\,701 = 73$ MPa
Axial stress $= 30\,000/1079 = 27$ MPa
Total stress $= 73 + 27 = 100$ MPa

Input the following data in the 'Program Euro B for Cold Rolled Sections', which is on the CD-ROM under the 'Program for Calculating Allowable Compressive Stress versus Slenderness Ratio':

Yield strength, $Y = 210$
Outer diameter, $D = 88.9$ mm
Thickness, $t = 4.05$
$K = 0.85$
Length, $l = 2800$ mm

From the output, $\omega_{buc} = 0.7255263$, the allowable compression is $0.7255263 \times 210 =$ 152 MPa.

The actual stress of 100 MPa is less than the value of 150 MPa printed out by the computer. Hence a tube of 88.9 mm × 4.05 mm is suitable.

## 2.7. Design of node connectors

### *2.7.1. Types of connector*

Broadly speaking, there are plate connectors and spherical connectors. We will illustrate the design of each by means of examples.

### *2.7.2. Design principles of Octatube plate connectors*

The plate connector essentially consists of two semi-octagonal plates with drilled holes placed at right angles and welded together (Fig. 2.2). The tubes meeting at a node are flattened at their ends and connected to the plate connector by means of bolts.

The three structural elements involved at a joint are:

- the member, which is either a flattened circular hollow section or a rectangular hollow section with a plate welded to it
- the Octatube plate connector
- Bolts of 5.6, 8.8 or 10.9 grade according to the BS 5950: Part 1 designation.

For a general joint, there are tension members, where

$F_{11}$ is the yield strength of the flattened tube
$F_{12}$ is the bearing strength of the bolt on the flattened tube
$F_{13}$ is the bearing strength of the bolt on the node plate
$F_{14}$ is the yield strength of the node plate
$F_{15}$ is the shear strength of the bolts

and compression members, where

$F_{21}$ is the bearing strength of the bolt on the flattened tube
$F_{22}$ is the bearing strength of the bolt on the node plate
$F_{23}$ is the shear strength of the bolts.

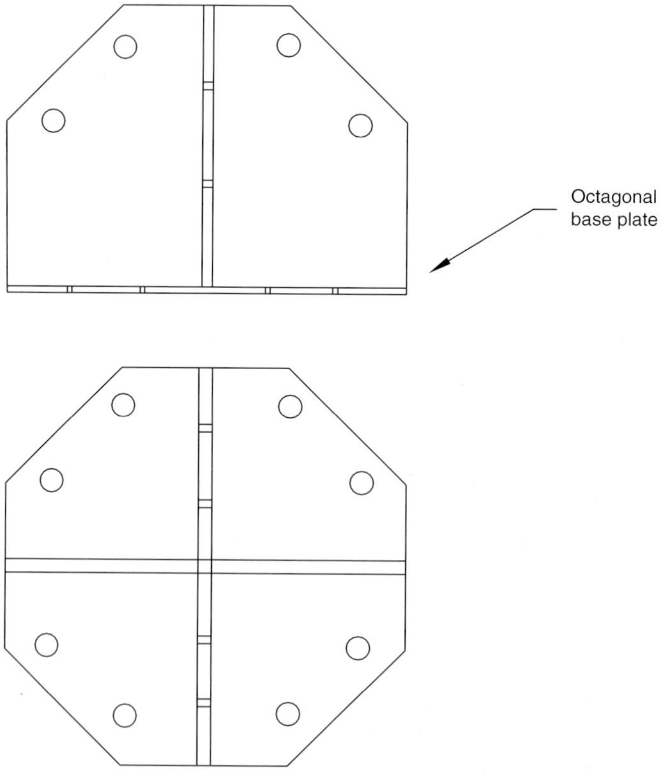


*Fig. 2.2. The Octatube plate connector*

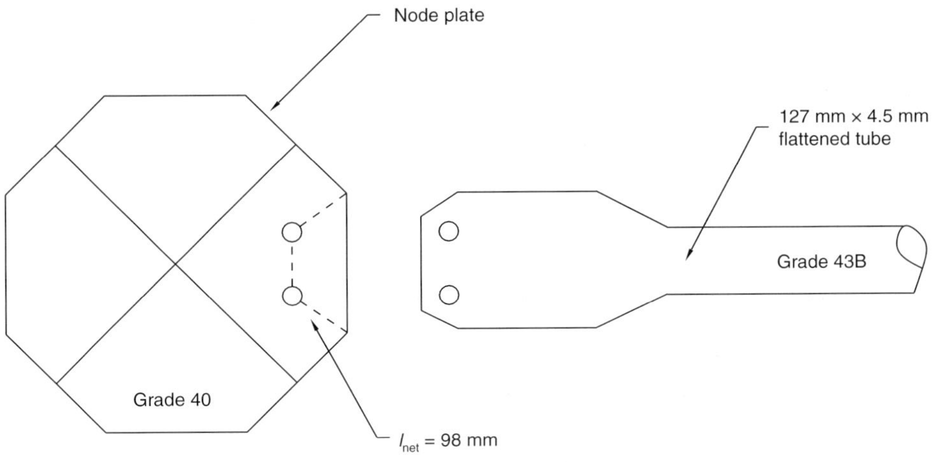


*Fig. 2.3. Details of the joint in Example 2.2*

The joint is safe if the force $F$ transmitted by the member to the joint satisfies the relation

$$|F| \leq F_{ij} \tag{2.14}$$

where $i = 1$ for tension and $i = 2$ for compression, $j = 1, 2, 3, 4, 5$ for tension and $j = 1, 2, 3$ for compression.

The example that follows illustrates the procedure for checking the safety of joints.

### *Example 2.2*

A flattened tube 127 mm × 4.5 mm is connected to an Octatube plate connector by means of two bolts of M20 in 21 mm diameter holes, the tubes having a yield strength of 265 MPa. The node plate is 8 mm thick and has a yield strength of 235 MPa. The effective area of the bolts is 245 mm$^2$ and the allowable stress $P_s$ in shear is 375 MPa. The bearing strength $P_{bs}$ is 460 MPa for the tube material and 374 MPa for the node material. Compute the strength of the joint.

Refer to Fig. 2.3.

(1) Yield strength of the flattened tube = (net area of flattened tube) × (yield strength)
Net area = [1733 – (2 × 2 × 21 × 4.5)] = 1355 mm$^2$
Yield strength = 265 MPa
Yield strength of flattened tube = (265 × 1355)/1000 = 359.08 kN.

(2) Yield strength of the node plate = (net area of node plate) × (yield strength)
Net area = (net length) × (thickness) × (No. of plates)
= 98 × 8 × 1 = 784 mm$^2$
Yield strength = 235 MPa
Yield strength of node plate = (784 × 235)/1000 = 184.24 kN.

(3) Shear strength of the bolts = (number of bolts) × (number of shear planes) × (area) × (allowable stress)
$= (2 \times 1 \times 245 \times 375)/1000 = 183.75$ kN.

(4) Bearing strength of the bolts on the flattened tube = (No. of plates) × (No. of bolts) × (diameter of bolts) × (thickness of tube) × (permissible stress)
$= (2 \times 2 \times 20 \times 4.5 \times 460)/1000 = 165.60$ kN

(5) Bearing strength of the bolts on the node plate = (No. of plates) × (No. of bolts) × (diameter of bolts) × (thickness of plate) × (permissible stress)
$= (1 \times 2 \times 20 \times 8 \times 374)/1000 = 119.68$ kN.

Hence, the joint can safely carry a tension or compression of 119.68 kN.

## 2.8. Design principles of Tuball spherical connectors

The strength of a joint is dependent on:

- the strength of the node connector
- the strength of the bolt
- the strength of the weld connecting the solid prop to the tube.

For a connection to be safe none of these strengths should exceed the force transmitted by the member to the joint.

*Example 2.3*

A tubular member of a space frame 60.3 mm × 2.9 mm carries a tension of 100 kN at the ultimate limit state. Design its connection to the spherical Tuball node. The allowable stress in weld is 375 MPa, the yield stress of the prop material $\sigma_e$ is 375 MPa, the proof stress $\sigma_{0.2}$ of the grade 10.9 bolt material is 940 MPa.

*Design of the bolt*: try an M16 bolt with an effective area of 157 mm$^2$.

Strength of the bolt = $(157 \times 940)/1000 = 147.58$ kN

This value is above the tension of 100 kN, and thus the bolt is suitable.

*Design of the solid prop*:

$$l > 0.8d\left(\frac{\sigma_{0.2}}{\sigma_e}\right)\left(\frac{N}{N_{bolt}}\right)$$

where $l$ is the theoretical length of the prop, $d$ is the diameter of the bolt, $\sigma_{0.2}$ is the proof strength of the bolt material, $\sigma_e$ is the yield strength of the prop material and $N_{bolt}$ is the strength of the bolt. Substituting the given values,

$$l > (0.8 \times 16 \times 940 \times 100)/(375 \times 147.58)$$
$$l > 21.7 \text{ mm}$$

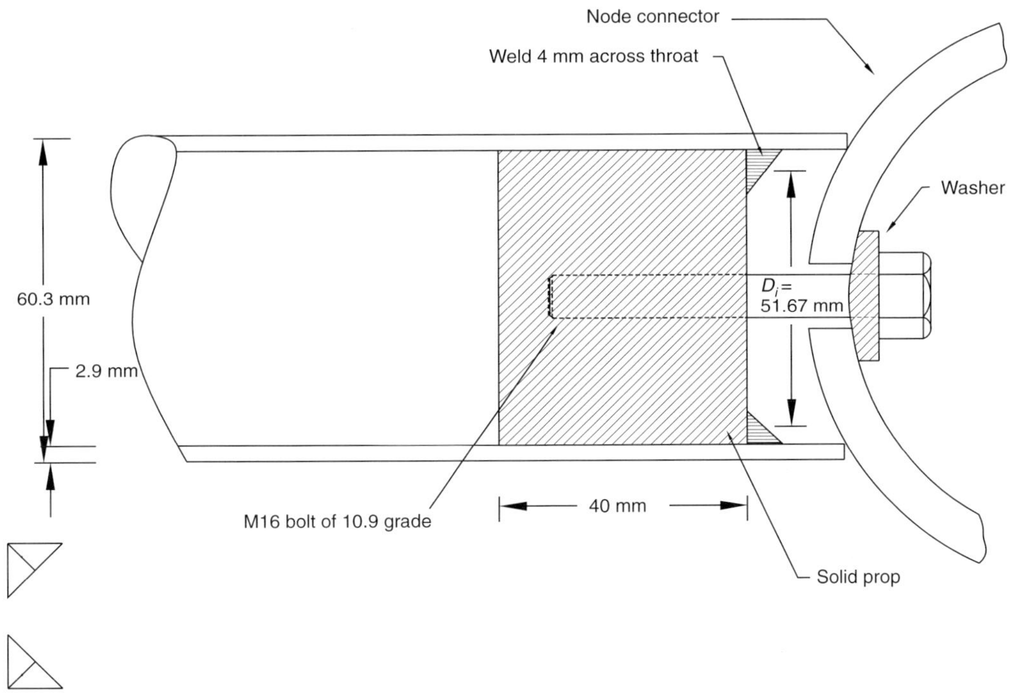


*Fig. 2.4. Detail of the connections in Example 2.3*

The practical length is 40 mm (Fig. 2.4).

*Design of the weld*:

Strength of weld = (length of weld) × (throat dimension) × (permissible stress)

Referring to Fig. 2.4, the length of the weld is

$$\pi D_i = \pi(60.3 - 2 \times 2.90 - 2.83) = 153.44 \text{ mm}$$

Thus the strength of weld is

$$(162.32 \times 4 \times 3.75)/1000 = 243 \text{ kN}$$

This value is above the tension of 100 kN, and thus the weld is suitable.

We will use a 135 mm diameter node which has a breaking strength of 395 kN. With a factor of safety of 2.0, the node can safely carry 197.50 kN (i.e. > 100 kN).

## 2.9. Properties of bolts

The effective areas of bolts are given in Table 2.2. The mechanical properties of grade 8.8 and 10.9 bolts given in Table 2.3 are typical values for these classes of bolt.

*Table 2.2. The effective areas of bolts*

| Threaded size | Effective area (mm$^2$) |
|---|---|
| M4 | 8.78 |
| M5 | 14.2 |
| M6 | 20.1 |
| M8 | 36.6 |
| M10 | 58.0 |
| M12 | 84.3 |
| M14 | 115.0 |
| M16 | 157.0 |
| M18 | 192.0 |
| M20 | 245.0 |
| M22 | 303.0 |
| M24 | 353.0 |
| M27 | 459.0 |
| M30 | 561.0 |
| M33 | 694.0 |
| M36 | 817.0 |
| M39 | 976.0 |
| M42 | 1120.0 |

*Table 2.3. The mechanical properties of grade 8.8 and 10.9 bolts*[a]

| Property | Grade 8.8, ≤ M16 (N/mm$^2$) | Grade 8.8, > M16 (N/mm$^2$) | Grade 10.9, all diameters (N/mm$^2$) |
|---|---|---|---|
| Tensile strength, minimum | 800 | 830 | 1040 |
| Yield strength, 0.2% offset minimum | 640 | 660 | 940 |
| Proof load stress | 580 | 600 | 830 |
| Shear strength, minimum | 480 | 498 | 624 |
| Hardness: Brinell HB | 219–285 | 242–319 | 295–362 |
| Hardness: Rockwell HRC | 20–30 | 23–34 | 31–39 |
| Elongation (% on GL)[b] | 12% minimum | 12% minimum | 9% minimum |

[a]From the brochure of Precision Fasteners Ltd, Bombay, in collaboration with SPS Technologies Inc., USA

[b]Elongation (% on GL) = $\sqrt{5.65\,A}$, where $A$ is the cross-sectional area

## 2.10. Coning of tubes at nodes

The diameter of the node connector is selected by taking into account the number and the diameters of the tubes meeting at a node. It may sometimes happen that the space between members on the connector surface is too small. To relieve congestion, the tubes may be coned (see Fig. 1.5).

## 2.11. Use of a threaded rod

If the force transmitted by a member exceeds the safe strength of the connector, the tube member may be terminated at the connector and a threaded rod run through the connector and connected to the tube at the other end (see Fig. 1.5). The node connector, being an elastic member, will also carry a small part of the load. The sharing of the load between the threaded rod and the node connector can be worked out according to the ratio of their stiffnesses.

## 2.12. Design of purlins

Sections of purlins may be arrived at by using the empirical procedure given in BS 5950: Part 1, Clause 4-12-4-3, Table 2-9.

*Example 2.4*

The Tuball connectors of a space frame are located on a grid of 3 m × 3 m. Design a purlin, given that the loading is 125 kg/m$^2$.

Using the notation of BS 5950: Part 1,

Load, $W_p = (3 \times 3)(125) = 1125$ kg = 11.03 kN
$l = 3000$ mm
Modulus of section, $Z = W_p\ (l/100) = 11.03/(3000/1800) = 18.38$ cm$^3$

A cold-formed channel 100 mm × 50 mm × 4 mm with $Z = 21.096$ cm$^3$ will suffice.

## 2.13. Codes of practice

A search of the literature revealed that the only effort ever made in the past to provide some broad guidelines to the designer of space frames was the report of Working Group 8 of the International association of Shell and Spatial Structures [2.3]. But this report, entitled *Analysis, Design and Realization of Space Frames*, is not written in the format of a code. In the absence of national or international codes specific to space frames, general national codes applicable to all steel structures are used. The only important difference between national codes is that they gave different formulae for the design of compression members. Surprisingly, these seemingly different formulae give results that agree so closely that they plot one over the other. This is demonstrated in Fig. 2.5 in which the allowable compressive stresses corresponding to different slenderness ratios, as specified by Eurocode 3: 1989, BS 5950: Part 1: 1985, the Indian code (IS 800: 1984) and the Australian (AS 4100-1990) and American codes, are compared for a yield strength of 240 MPa. The graph clearly shows that space frame design is not very code dependent.

## 2.14. Case study: a space frame with plate connectors for supporting a tennis court at Deira City centre, United Arab Emirates

The consultants for this project were Schuster Pechtold & Partners, Dubai, the design architects were The Callison Partnership Ltd, Seattle, USA, and the contractors were Galadari Engineering Works, Dubai.

### 2.14.1. Scope of the project

The project involved the design and construction of a flat double-layer space frame to support a concrete tennis court. The concrete, which was poured into steel deck sheets, rests on a tubular space frame, of rectangular plan, measuring 58.8 m × 50.4 m. The structure is supported on peripheral columns, spaced 8.40 m apart, and four interior latticed columns by means of inverted pyramidal 'tree' supports, which are a part of the space frame (Figs 2.6 and Fig 2.7).

### 2.14.2. Design brief

#### 2.14.2.1. Loads

The space frame is to be designed to carry the following loads:

- dead loads from concrete and steel deck 150 kg/m$^2$
- estimated weight of installations 50 kg/m$^2$
- weight of space frame, purlins, nodes, stubs and stools 50 kg/m$^2$
- live load 50 kg/m$^2$.

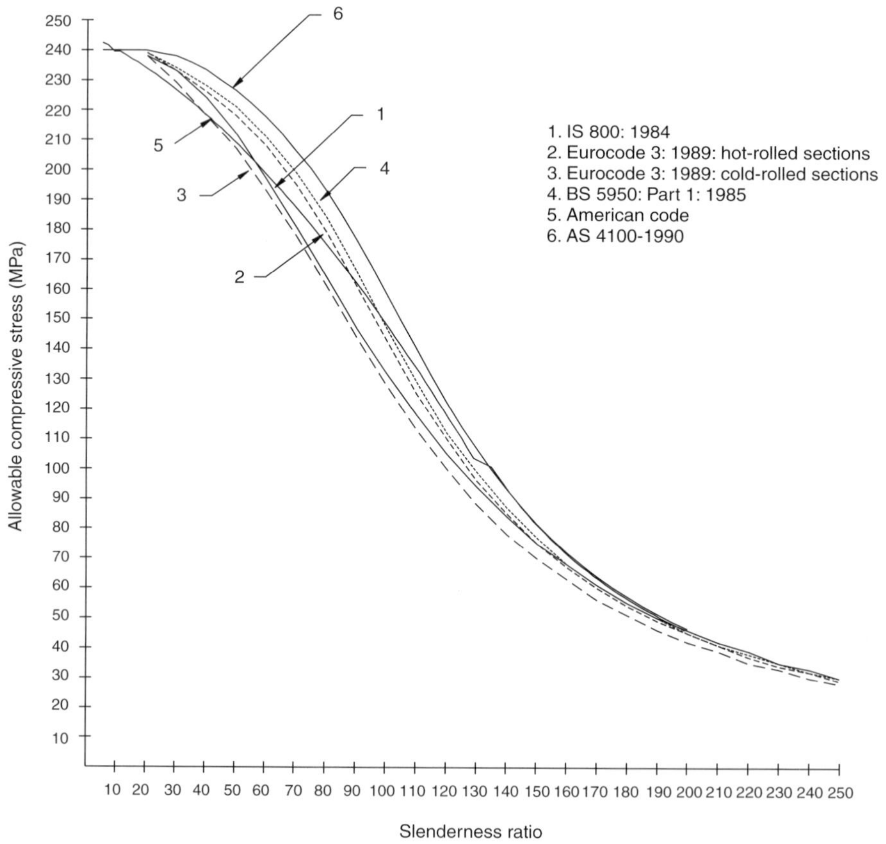


*Fig. 2.5. Comparison of the allowable compressive stresses (for a yield strength of 240 MPa) corresponding to different slenderness ratios, as specified by various codes*

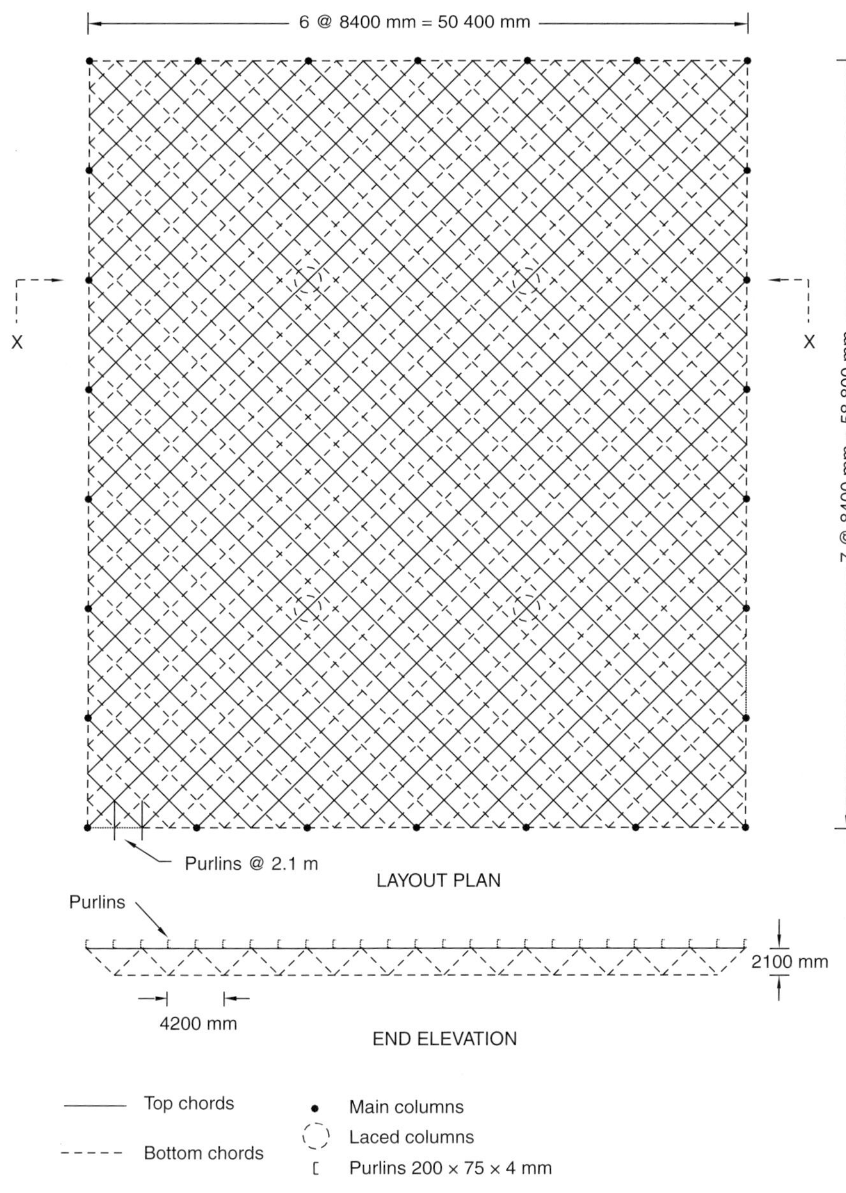


*Fig. 2.6. The space frame for the tennis court at Deira City centre*

Total load: 380 kg/m$^2$ = 3.80 kN/m$^2$.
Load at ultimate limit state: 1.5 × 3.80 = 5.70 kN/m$^2$.

### *2.14.2.2. Code of practice*

Design to be based on the method of limit states. Design is generally to conform to BS 5950: Part 1: 1985. Compression members are to be designed in accordance with Eurocode 3/BS 5950: Part 1: 1985 code provisions.

#### 2.14.2.3. *Deflections*

Deflections under live loads are to be less than 20 mm.

#### 2.14.2.4. *Specifications of tubes*

Tubes of mild steel are to conform to grade 43B of the BS 5950: Part 1: 1985 specifications. The yield strength is 265 MPa.

#### 2.14.2.5. *Node connectors*

Octatube plate connectors are specified.

### 2.14.3. *Design decisions*

#### 2.14.3.1. *Choice of topology*

When we received this rather unusual mandate from Dubai to design a space frame to support a concrete tennis court, our immediate reaction was to contact Galadari Engineering Works at Dubai to make sure if the client needed this space frame to carry the tennis court or to cover it! When Galadari confirmed that the idea was to play tennis on the top of the space frame, we were faced with the formidable challenge posed by the abnormally heavy loading and the stringent restriction on deflections under live load. It was quite clear from the very beginning that, to win the contract against stiff global competition, the key to meeting the two conflicting objectives of minimizing weight while ensuring adequate stiffness to keep deflections under control lay in the judicious selection of the appropriate topology. The six topologies normally used are shown in Fig. 2.8. We opted for the very stiff and structurally efficient topology of square on square, set diagonally (type 2).

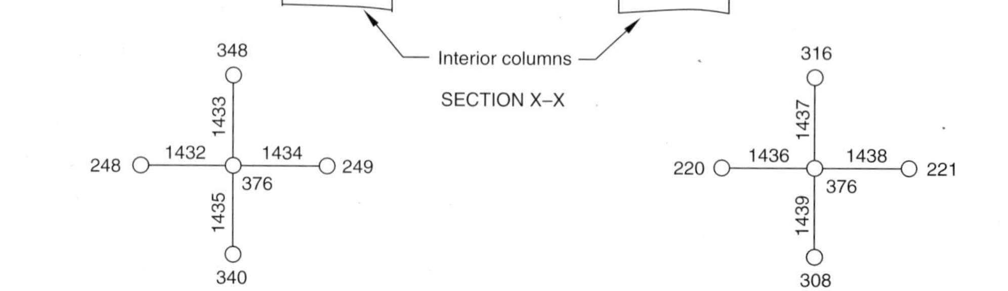


*Fig. 2.7. Plan of the pyramidal columns for the tennis court at Deira City centre*

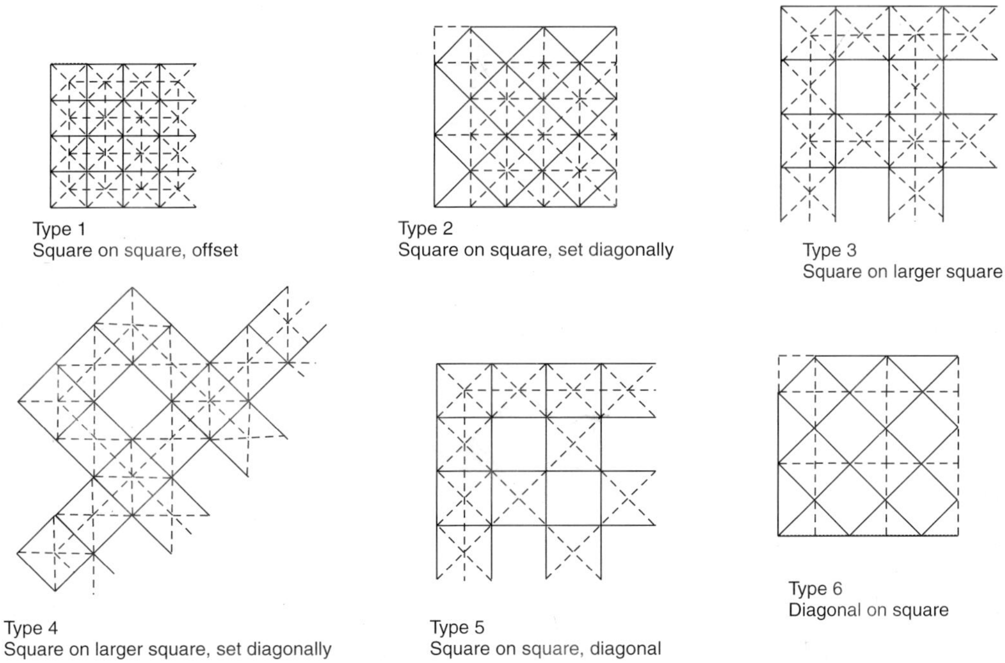


*Fig. 2.8. The six grid arrangements normally used for space frames*

### 2.14.3.2. Choice of grid

The choice of a grid spacing of 2.97 m in the diagonal direction permitted purlins to be spaced 2.1 m apart (see Fig. 2.6).

### 2.14.3.3. Selection of structural depth

A structural depth of 2.1 m was chosen because it makes the length of the diagonal bracing members equal to the diagonally laid out top and bottom chords. The layout of the top and bottom chords and the node and the member numbering schemes adopted are shown in Fig. 2.9.

## 2.14.4. Structural analysis and optimization

Taking advantage of symmetry, half the space frame was analysed using SAP90 software. Based on previous experience, a tube size of 127.0 m × 4.50 m was adopted for the top and bottom chords for starting the first cycle of analysis. Similarly, for diagonal bracing members, a size of 76.10 m × 3.25 m was selected. The forces in the members found from the first cycle of analysis were scanned, and tube sizes to match the forces were selected and new input data written before the second cycle was commenced. This iterative process was continued until the difference in weight of the space frame resulting from two successive cycles was insignificant. For all practical purposes, this iterative process optimizes the structure to minimum weight. An automatic iterative procedure, using an expert system, to arrive at the minimum weight is presented in Chapter 9.

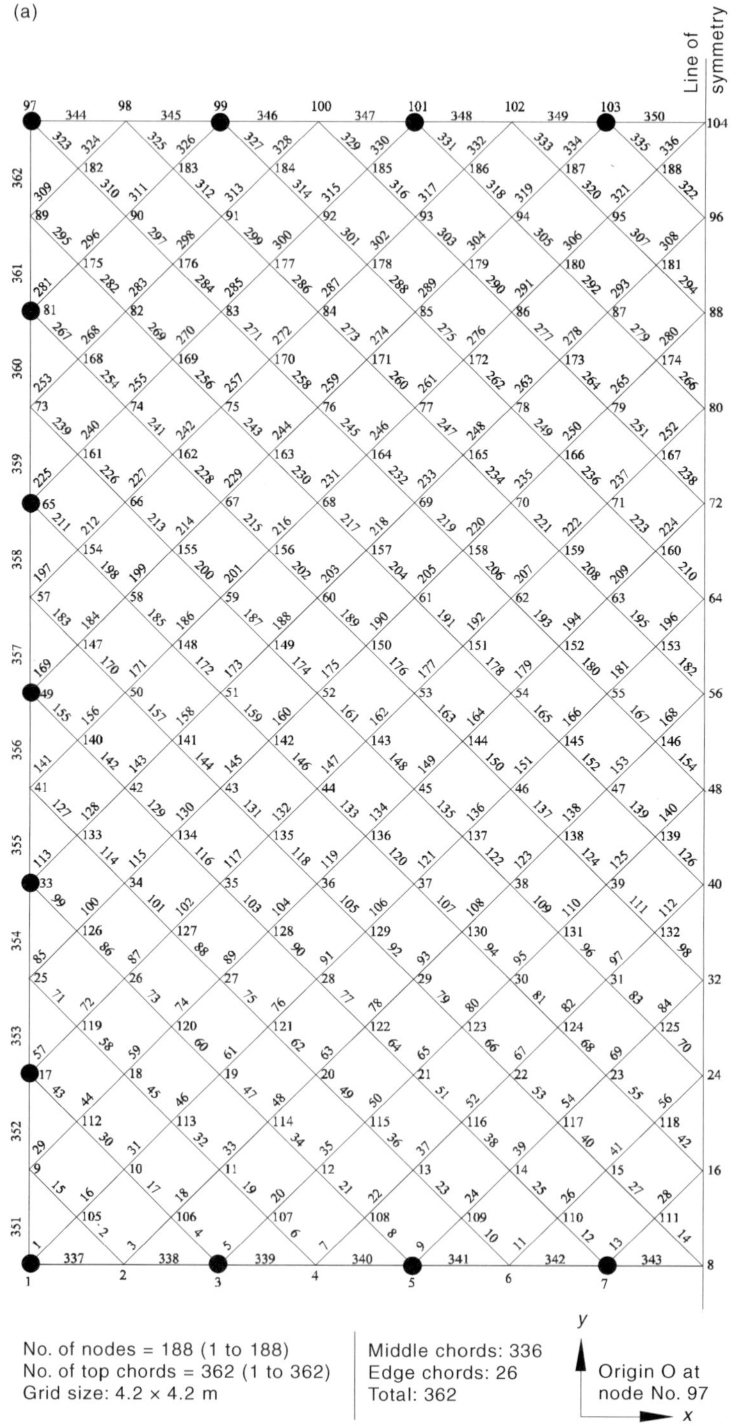


*Fig. 2.9. Half-plans, showing node and member numbering: (a) top chords*

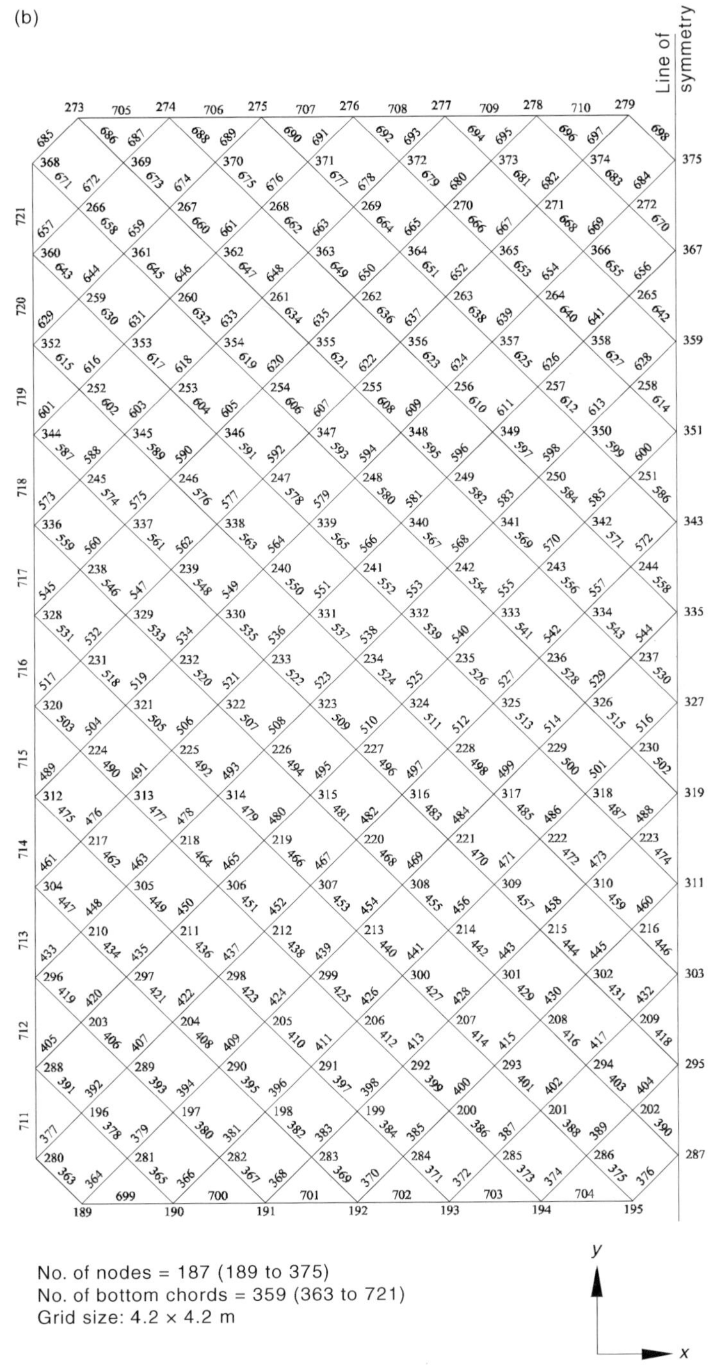


*Fig. 2.9 (Contd). (b) Bottom chords*

(c)

No. of diagonal bracings = 710 (722 to 1431)
No. of top nodes = 188 (1 to 188)
No. of bottom nodes = 187 (189 to 375)

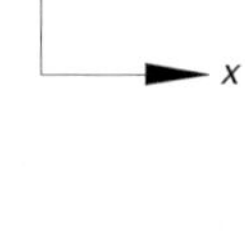


*Fig. 2.9 (Contd). (c) Diagonal bracing*

*Table 2.4. The allowable tension and compression in tubes at the ultimate limit state*

| Tube size | Area (mm) | $r$ (mm) | $l_e/r$ | Permissible compression, BS 5950: Part 1: 1985 (kN) | Permissible tension (kN) |
|---|---|---|---|---|---|
| *Top chords, bottom chords and bracings*[a] | | | | | |
| 76.20 × 3.25 | 744 | 25.76 | 98 | 118 | 197 |
| 114.30 × 4.50 | 1553 | 38.82 | 65 | 350 | 412 |
| 127.00 × 4.50 | 1733 | 43.31 | 58 | 406 | 459 |
| 139.70 × 4.50 | 2277 | 47.48 | 53.17 | 546 | 683 |
| 193.70 × 5.40 | 3196 | 66.57 | 38 | 805 | 847 |
| 219.10 × 5.90 | 3953 | 75.38 | 33 | 1011 | 1047 |
| *Edge chords*[b] | | | | | |
| 127.00 × 4.50 | 1733 | 43.31 | 82 | 336 | 459 |
| 88.90 × 4.05 | 1080 | 30 | 119 | 130 | 286 |

[a]Length = 2970 mm; effective length, $l_e = 0.85 \times 2970 = 2524.5$ mm; yield strength = 265 MPa
[b]Length = 4200 mm; effective length, $l_e = 0.85 \times 4200 = 3570$ mm; yield strength = 265 MPa

### 2.14.5. Computer input and output

The input data (the sizes of the members and the boundary conditions assumed over the columns) and the computer output (in the form of member forces, displacements and reactions over the supports) are on the accompanying CD-ROM.

### 2.14.6. Design summary

#### 2.14.6.1. Top and bottom chords

The maximum tension of 473.11 kN was found to occur in members 121 and 219. The maximum compression of 286.41 kN occurred in members 55 and 293. The allowable tension and compression forces in tubes of different sizes are presented in Table 2.4. The allowable compression was computed using the data in BS 5950: Part I, Table 27(a). It can be seen from Table 2.4 that a size of 139.70 mm × 5.40 mm is adequate for members 121 and 219. To facilitate detailing at nodes 69 and 37, where members 121 and 219 are located, a size of 139.70 mm × 5.40 mm was adopted for all top chords meeting at these nodes. For all other diagonally laid top chords, a tube of 127.0 mm × 4.50 mm was adopted. For edge chords, which carry only small forces, a tube of 88.90 mm × 4.05 mm was found to be adequate.

#### 2.14.6.2. Diagonal bracings

Dark lines (Fig. 2.10) indicate the zones of heavy shear around the interior columns. In this region, a tube of 114.30 mm × 4.50 mm is proposed for all diagonal bracing members, except 840, 843, 952 and 955. For these heavily stressed members a tube of 139.70 mm × 5.40 mm with a buckling strength of 537 kN will be adequate.

### 2.14.6.3. Pyramidal support members over the interior columns

From the computer output it can be seen that the maximum compression in the sloping diagonals forming the pyramidal support over the interior columns is 930.11 kN, which requires a tube size of 219.10 mm × 5.90 mm with a buckling load of 1004 kN.

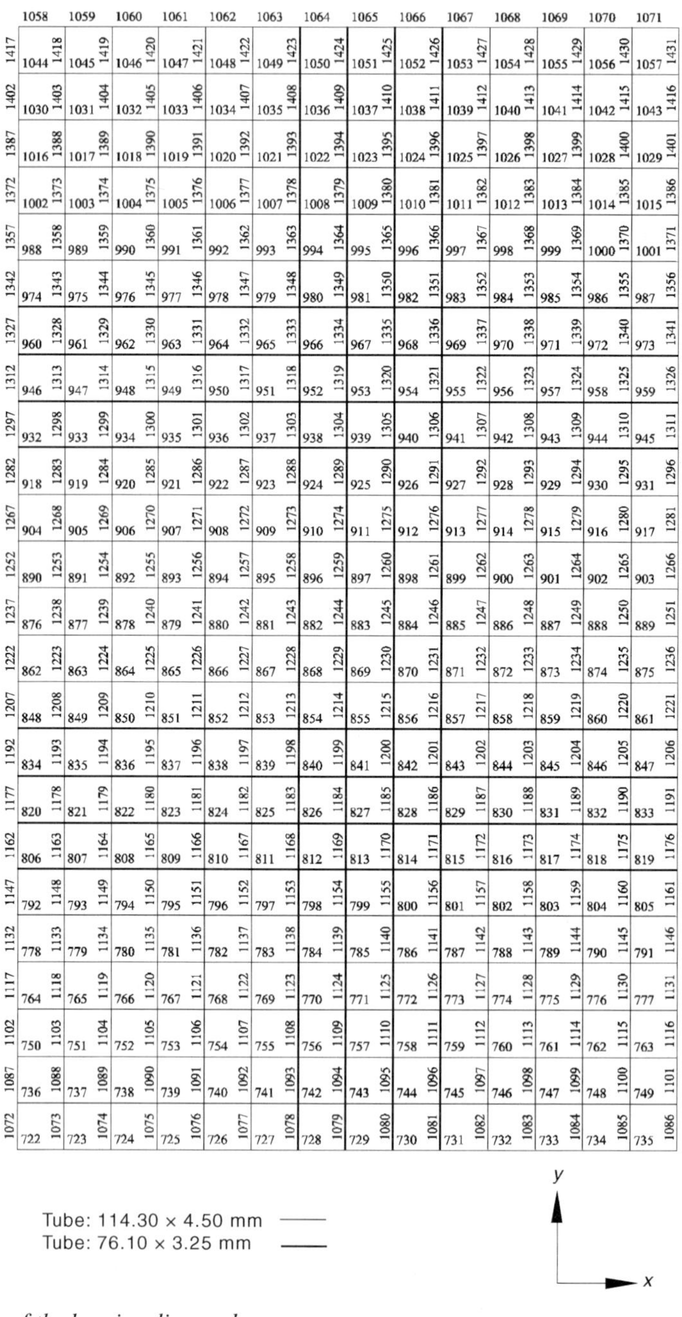


*Fig. 2.10. Sizes of the bracing diagonals*

### *2.14.6.4. Check for deflections*

The maximum deflection of 65.78 mm occurs at node 56 at the ultimate limit state. The corresponding deflection at the serviceability limit state is arrived at as $65.78/1.50 = 43.85$ mm. The deflection attributable to live load works out as $(1.5/3.8) \times 43.85 = 17.31$ mm, which is within the prescribed limit of 20 mm.

### *2.14.6.5. Support reactions*

The reactions at the ultimate limit state found from the computer output are given in Table 2.5.

## 2.15. Design of plate connectors

The general principles in the design of plate connectors are as follows:

- Each connection will involve the use of at least two bolts.
- When necessary, the ends of the flattened tubes will be welded to a flat plate, either to compensate for the loss of cross-section resulting from the deduction of bolt holes or to accommodate more bolts, especially when more than two are used. Alternatively, the end of the tube may be slotted for an adequate length and a plate inserted in the slot and welded. The plate so welded will be connected to the node plate by means of bolts. This arrangement will permit the use of more then two bolts.
- Generally, single shear connections are used for diagonal bracings.
- For top and bottom chords, double plates with bolts in double shear will be adopted.

### *2.15.1. Computation of the strength of joints*

The design principles of Octatube plate connectors have already been dealt with in detail in Section 2.7, with aid of an illustrative example, outlining the step-by-step procedure involved in computing the strength of a joint.

### *2.15.2. The plate connector used in Deira City centre*

The details of the Octatube connector used in this project are shown in Fig. 2.11. The yield stress of the node plate of grade 40 was 235 MPa. Bolts of 20 mm diameter and grade 8.8 were used in 21 mm holes. The effective area of the bolts was 245 mm$^2$. The shear strength of the bolts at the ultimate limit state is 375 MPa. The strength in bearing of the node plate of grade 40 was 374 MPa, and that of the tubes of grade 43 was 460 MPa.

### *2.15.3. Design of a typical joint*

By way of illustration, we present the design of the joint at node 69. At this joint there are four top chord members (232, 233, 218 and 219) and four diagonal bracing members (953, 954, 1305 and 1320) (Fig. 2.12).

The top chords at node 69 are:

- 232, connecting 69 to 164
- 233, connecting 69 to 165
- 218, connecting 69 to 157
- 219, connecting 69 to 158.

*Table 2.5. The support reactions at the ultimate limit state*

| Joint | $F(x)$ | $F(y)$ | $F(z)$ |
|---|---|---|---|
| 1 | 0.0000 | 0.0000 | 41.2201 |
| 3 | 43.5833 | 0.0000 | 284.7296 |
| 5 | 24.4813 | 0.0000 | 199.5754 |
| 7 | –39.6630 | 0.0000 | 396.0137 |
| 8 | –62.4239 | 0.0000 | 0.0000 |
| 16 | –319.3435 | 0.0000 | –25.1400 |
| 17 | 0.0000 | 1.0963 | 327.6540 |
| 24 | –351.0956 | 0.0000 | –25.1400 |
| 32 | –285.7508 | 0.0000 | –25.1400 |
| 33 | 0.0000 | 45.3238 | 232.9930 |
| 40 | –229.0011 | 0.0000 | –25.1400 |
| 48 | –284.4043 | 0.0000 | –25.1400 |
| 49 | 0.0000 | 0.0000 | 250.6006 |
| 56 | –347.5121 | 0.0000 | –25.1400 |
| 64 | –284.4043 | 0.0000 | –25.1400 |
| 65 | 0.0000 | –45.3238 | 232.9930 |
| 72 | –229.0011 | 0.0000 | –25.1400 |
| 80 | –285.7508 | 0.0000 | –25.1400 |
| 81 | 0.0000 | –1.0963 | 327.6540 |
| 88 | –351.0956 | 0.0000 | –25.1400 |
| 96 | –319.3435 | 0.0000 | –25.1400 |
| 97 | 0.0000 | 0.0000 | 41.2201 |
| 99 | 43.5833 | 0.0000 | 284.7296 |
| 101 | 24.4813 | 0.0000 | 199.5754 |
| 103 | –39.6630 | 0.0000 | 396.0137 |
| 104 | –62.4239 | 0.0000 | 0.0000 |
| 287 | 243.1125 | 0.0000 | 0.0000 |
| 295 | 342.9328 | 0.0000 | 0.0000 |
| 303 | 314.9769 | 0.0000 | 0.0000 |
| 311 | 238.9729 | 0.0000 | 0.0000 |
| 319 | 236.0532 | 0.0000 | 0.0000 |
| 327 | 301.3255 | 0.0000 | 0.0000 |
| 335 | 301.3255 | 0.0000 | 0.0000 |
| 343 | 236.0532 | 0.0000 | 0.0000 |
| 351 | 238.9729 | 0.0000 | 0.0000 |
| 359 | 314.9769 | 0.0000 | 0.0000 |
| 367 | 342.9328 | 0.0000 | 0.0000 |
| 375 | 243.1125 | 0.0000 | 0.0000 |
| 376 | 0.0000 | 89.0691 | 2451.8738 |
| 377 | 0.0000 | –89.0691 | 2451.8738 |

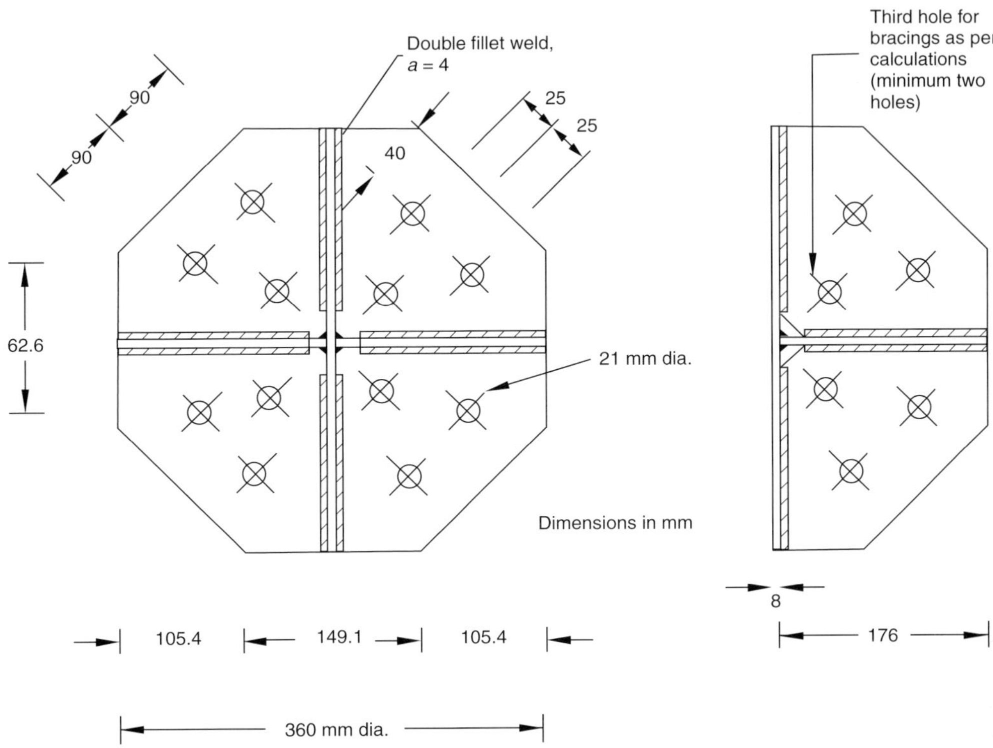


*Fig. 2.11. The Octatube connector used in the Deira City centre project*

The bracing diagonals at node 69 are:

- 953, connecting top node 69 to bottom node 248
- 954, connecting top node 69 to bottom node 249
- 1305, connecting top node 69 to bottom node 340
- 1320, connecting top node 69 to bottom node 348.

The forces in the eight members connected together at node 69 are given in Table 2.6.

### *2.15.3.1. Joint design for top chords at node 69*

The joint will be designed for the most highly stressed top chord member, in this case No. 219 carrying a compression of 473.11 kN. The same design will then be adopted for the other three top chords meeting at node 69.

The size of the four top chords meeting at node 69 is 139.70 mm × 5.40 mm. The tubes will be slotted centrally at the end for a distance of 200 mm, and a plate 170 mm wide and 15 mm thick inserted into the slot and welded to the tube (Fig. 2.13). An additional plate 8 mm thick is provided in the node.

The strength of the joint is calculated as follows.

*Tension*:
Net area of shear plate = $(170 \times 15) - (2 \times 15 \times 21) = 1920$ mm$^2$
Permissible tension in shear plate = $(1920 \times 265)/1000 = 508.8$ kN
Net area of node plate = $(2 \times 204 \times 8) = 3264$ mm$^2$
Permissible tension in node plate = $(235 \times 3264)/1000 = 767$ kN.

*Shear*:
Shear strength of four bolts in double shear = $(2 \times 4 \times 245 \times 375)/1000 = 735$ kN.

*Bearing*:
Bearing on shear plate = $(1 \times 4 \times 15 \times 20 \times 460)/1000 = 552$ kN
Bearing on node plates = $(2 \times 4 \times 8 \times 20 \times 374)/1000 = 478.72$ kN.

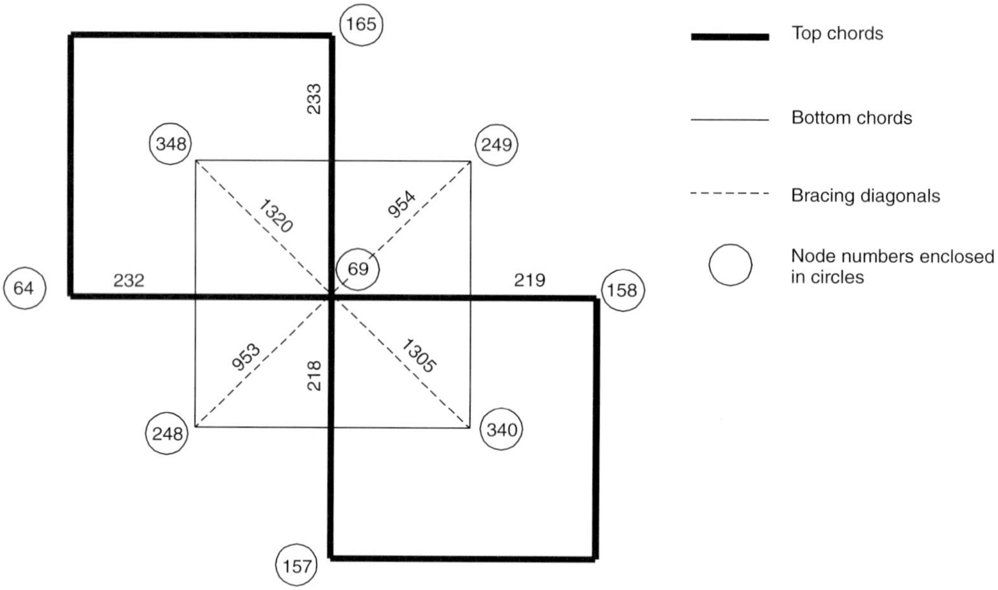


*Fig. 2.12. The joint at node 69*

*Table 2.6. The forces in the eight members connected at node 69*

| Member No. | Force (kN) |
|---|---|
| 218 | 431.11 |
| 219 | 473.11 |
| 232 | 454.39 |
| 233 | 412.42 |
| 953 | –8.35 |
| 954 | –8.38 |
| 1305 | –45.89 |
| 1320 | –8.47 |

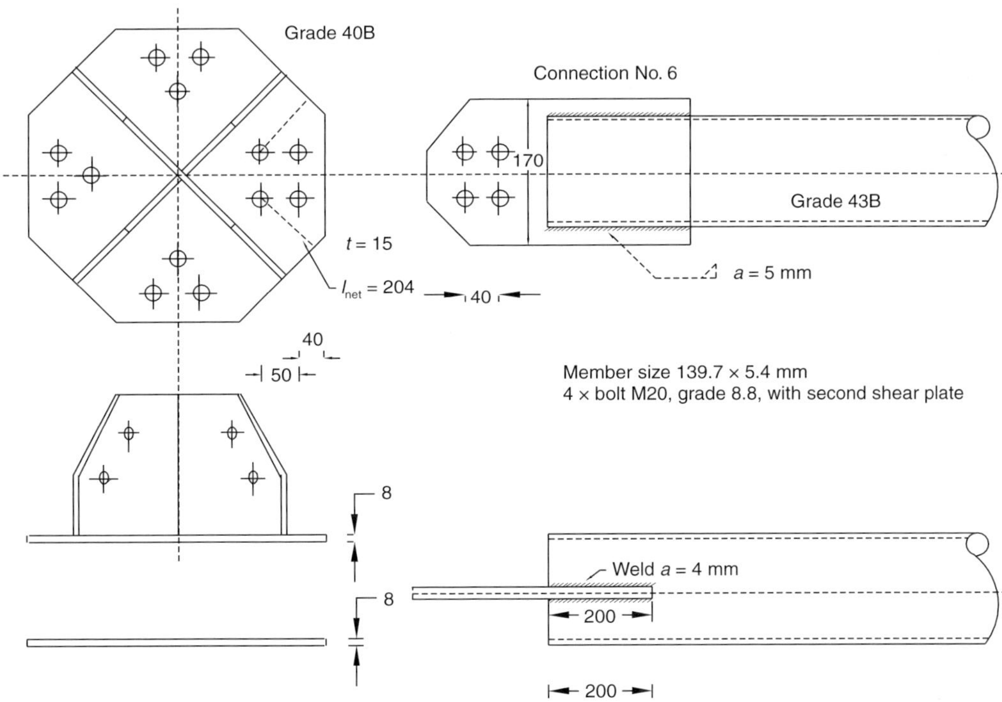


*Fig. 2.13. The top chords at node 69*

Hence the strength in the joint is 478.72 kN. This is above the compression force of 473.11 kN carried by the joint, and therefore the joint is safe.

The details of typical top chord members meeting at node 69 are shown in Fig. 2.14.

### *2.15.3.2. Joint design for diagonal bracings at node 69*

Although the forces in the bracing diagonals meeting at node 69 are small, these lie in the zone of high shear demarcated by the dark lines in Fig. 2.10. Hence, the size of the tubes used is 114.30 mm × 4.50 mm. The tube is flattened and connected directly to the node plate by means of two bolts (Fig. 2.15).

The strength of the joint is calculated as follows.

*Tension*:
Net area of flattened tube = $1553 - (2 \times 2 \times 4.5 \times 21) = 1175$ mm$^2$
Permissible tension in tube = $(1175 \times 265)/1000 = 311.375$ kN
Net area of node plate = $1 \times 98 \times 8 = 784$ mm$^2$
Permissible tension in node plate = $(784 \times 235)/1000 = 184.24$ kN.

*Shear* (bolts in single shear):
Safe shear of tube = $(1 \times 2 \times 245 \times 375)/1000 = 183.75$ kN.

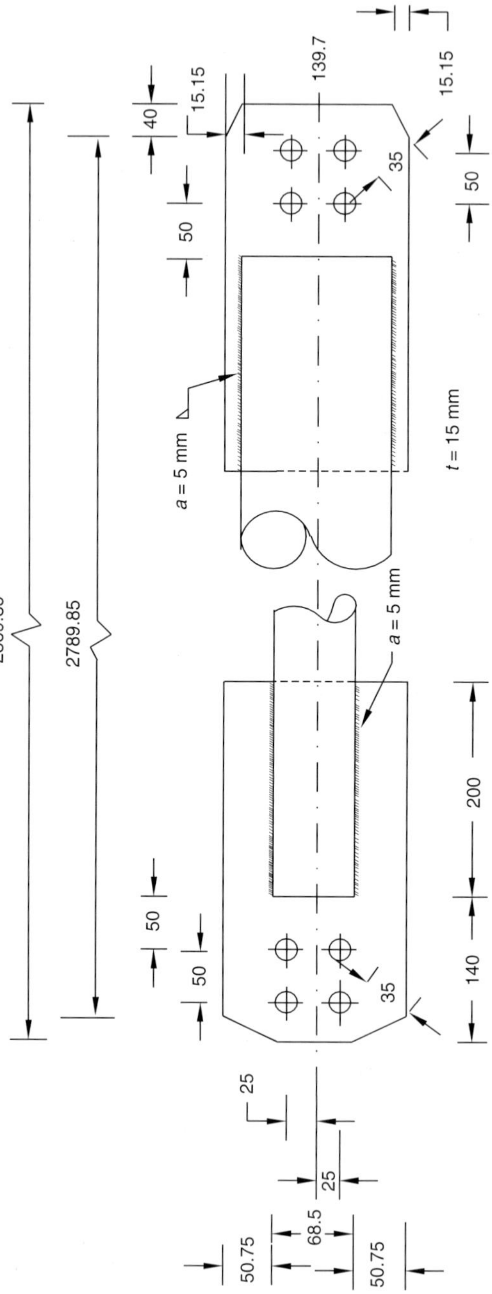


*Fig. 2.14. Details of the top chords at node 69*

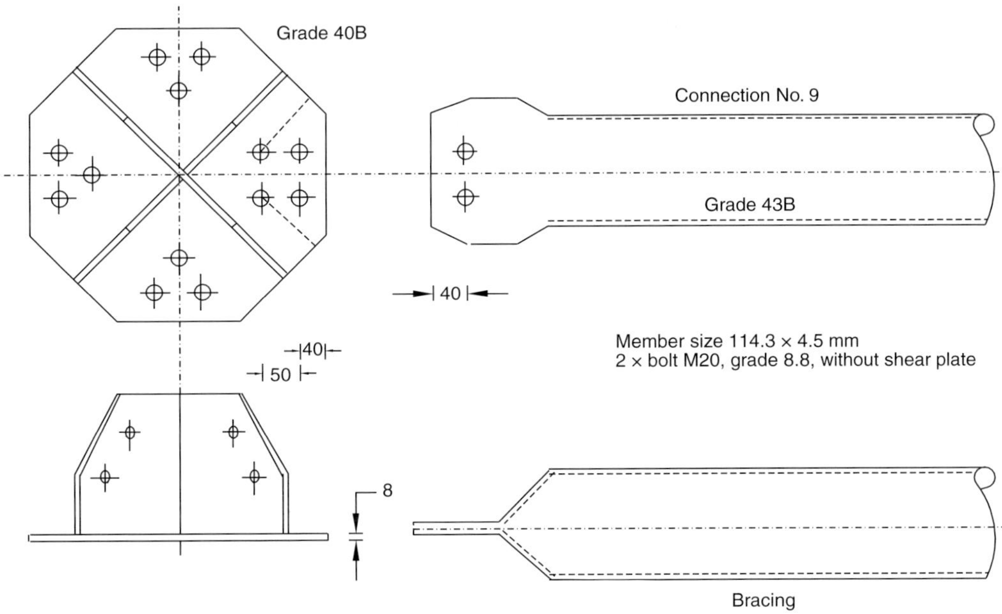


*Fig. 2.15. Details of the diagonal bracings at node 69*

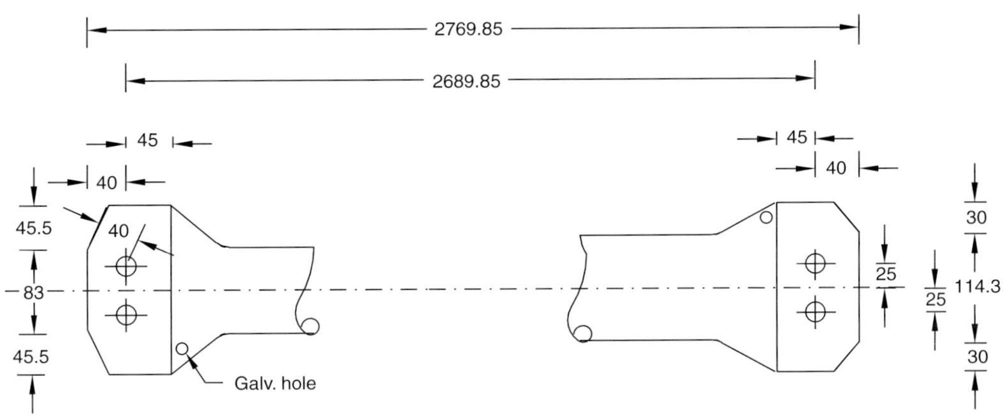


*Fig. 2.16. Details of a typical diagonal bracing member at node 69*

*Bearing*:
Bearing strength of tube = $(2 \times 2 \times 4.5 \times 20 \times 460)/1000 = 165.60$ kN
Bearing strength of node plates = $(1 \times 2 \times 8 \times 20 \times 374)/1000 = 119.68$ kN.

Hence the strength of the joint is 119.70 kN, which is adequate to transmit the forces in the bracing diagonals. Details of a typical diagonal bracing member at node 69 are shown in Fig. 2.16.

#### *2.15.3.3. Weight of the space frame*

The weight of the space frame is 89.50 t. The steel consumption works out to:

$$\frac{89\,500}{58.80 \times 50.40} = 30.20 \text{ kg/m}^2$$

Considering the unusually heavy loading for which the space frame is designed, the design may be regarded as structurally very efficient.

## 2.16. References

2.1. Eurocode No. 3. *Design of Steel Structures. Part 1: General Rules and Rules for Building*, 1989, vol. 1, ch. 1–9; vol. 2, Annexes.

2.2. BRITISH STANDARDS INSTITUTION. *BS 5950: Structural use of Steel Work in Building. Part 1: Code of Practice for Design of Simple and Continuous Construction: Hot Rolled Sections.* BSI, Milton Keynes, 1985.

2.3. Analysis, design and realization of space frames. A state-of-the-art report. *Bulletin of the International Association of Shell and Spatial Structures, Special Issue*, 1984, **15**(84/85).

# 3. Preliminary design

## 3.1. Getting started

It is often said that, while analysis is a science, design is very much an art. While analysis can be competently handled even by a novice, design often needs inputs from experienced experts, especially in the preliminary conceptual phases. To get started is the hardest part of all, especially if the structure is statically indeterminate. Unlike for statically determinate structures, for statically indeterminate structures the designer has to choose appropriate preliminary sizes of members before the first cycle of analysis can begin. Analogies of the kind presented in this chapter wherein the discrete space frame is replaced by its continuum equivalent are useful in estimating these sizes. These initial guess estimates need not be highly accurate. As the analysis proceeds through successive iterative cycles, the member sizes will quickly converge to the appropriate values, as explained in detail in Chapter 9. Information on the approximate sizes of members is also often required to estimate the weight of the space frame when bidding for jobs.

## 3.2. Continuum analogies

Space frames may be flat, singly curved reticulated barrel vaults or doubly curved reticulated domes. For the purposes of preliminary design, it is sufficiently accurate to replace the frames with their continuum equivalents. This chapter is concerned only with flat double-layer space frames, because they are repetitive structures of frequent occurrence.

The continuum equivalent of a double-layer space frame is a pair of laminae, separated by a distance $h$ equal to the structural depth of the space frame, as shown in Fig. 3.1. The top and bottom laminae, carrying only in-plane forces, replace the top and bottom chords of the space frame. The laminae represent an equivalent plate, the bending moments in which are easily found from the readily available formulae for plates. By dividing the bending moments in the equivalent plate by $h$, the maximum forces in the top and bottom chords and appropriate sizes of tubes to carry them can be worked out. Sizes so arrived at can be used to carry out the first cycle of structural analysis. The continuum will be statically equivalent to the space frame, when all the rigidity characteristics of both are identical. Kollar [3.1] has discussed conditions under which the equivalent continuum will be isotropic, orthotropic or aeolotropic. The substitution of space frames by continua has also been discussed by Heki and Saka [3.2], Renton [3.3, 3.4] and Soare [3.5], among others.

## 3.3. Continuum analogy for a double-layer space frame supported at the edges with square over square offset topology

It will be assumed that the sizes of the top and bottom chords are the same and that they are the same in both the $x$ and the $y$ directions. The bottom chords are, however, offset with respect to the top chords by half a grid spacing (Fig. 3.2).

The following notation will be used:

$A$, area of cross-section of the members
$E$, Young's modulus of elasticity
$a$, grid spacing measured parallel to co-ordinate axes
$h$, structural depth of the space frame
$w$, displacement in the $z$ direction
$p$, load per unit area
$D$, rigidity of the equivalent plate ($=AEh^2/2a$)
$l_y$, the longer linear plan dimension
$l_x$, the shorter linear plan dimension
$\alpha$, aspect ratio ($=l_y/l_x$)
$F_{max}$, maximum force in top/bottom chord

Renton [3.3, 3.4] has shown that, in the present case, the space frame may be replaced by the torsionless plate, the behaviour of which is governed by the following partial differential equation:

$$\frac{\partial^4 w}{\partial w^4} + \frac{\partial^4 w}{\partial y^4} = \frac{p}{D} \tag{3.1}$$

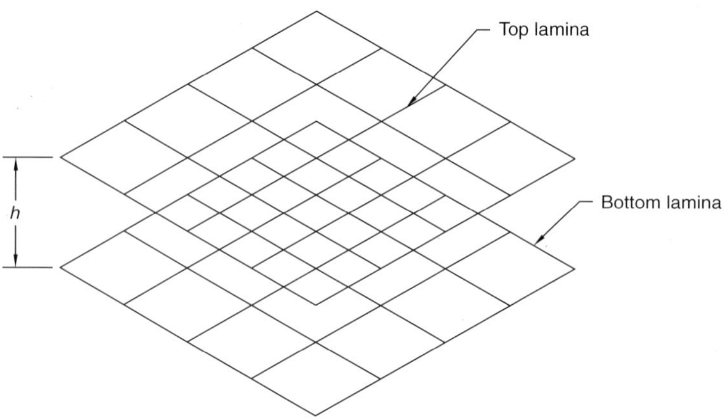


*Fig. 3.1. An equivalent plate*

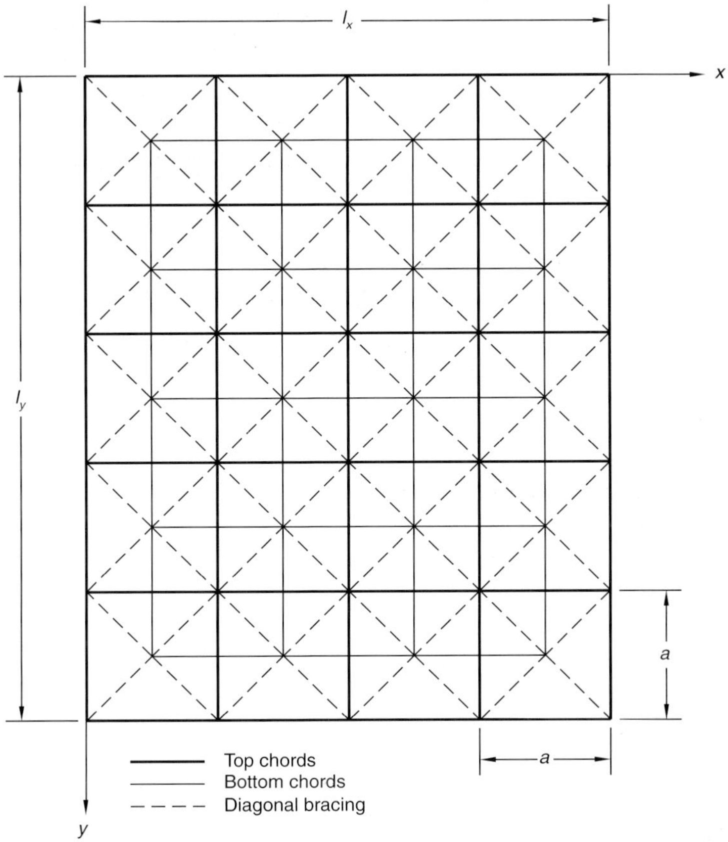


*Fig. 3.2. Square over square offset topology*

Hence,

$$w = \sum_{m=1,3,5,\ldots}^{\infty} \sum_{n=1,3,5,\ldots}^{\infty} \frac{16pl_x^4}{\pi^6 D} \frac{1}{1+1/\alpha^4} \sin\left(\frac{m\pi x}{l_x}\right) \sin\left(\frac{n\pi y}{l_y}\right) \tag{3.2}$$

As this series is rapidly convergent, we may restrict ourselves to the first term, and write

$$w_{\max} = \frac{16pl_x^4}{\pi^6 D} \frac{1}{1+1/\alpha^4} \tag{3.3}$$

Equation (3.3) may be rewritten as

$$w_{\max} = \frac{pl_x^4}{k_w D} \tag{3.4}$$

Restricting ourselves to the first term, and assuming Poisson's ratio $\nu = 0$, the expression for the maximum bending moment may be written as

*Table 3.1. Values of α, $k_w$, $k_M$ and $C_1$*[a]

| | | | | | | | |
|---|---|---|---|---|---|---|---|
| $\alpha$ | 1.0 | 1.1 | 1.2 | 1.3 | 1.4 | 1.5 | 1.6 |
| $k_w$ | 122.0 | 103.1 | 90.1 | 82.6 | 77.5 | 74.1 | 71.4 |
| $k_M$ | 13.1 | 10.9 | 9.6 | 8.7 | 8.2 | 7.8 | 7.5 |
| $C_1$ | 0.076 | 0.092 | 0.104 | 0.115 | 0.122 | 0.128 | 0.133 |

[a]Reproduced with permission from White and Gladischefski [3.6]

$$M_{\max} = \frac{16pl_x^2}{\pi^6 D}\frac{1}{1+1/\alpha^4} = \frac{pl_x^2}{k_M} \tag{3.5}$$

It is also possible to rewrite the above expression as

$$M_{\max} = C_1 pl_x^2 \tag{3.6}$$

where $C_1 = 1/k_M$. It is easy to verify that the maximum compression/tension is

$$F_{\max} = C_1 pl_x^2\left(\frac{a}{h}\right) = \frac{pl_x^2}{k_M}\left(\frac{a}{h}\right) \tag{3.7}$$

The coefficients $k_w$ and $k_M$ are given in Table 3.1. The use of the data given in Table 3.1 to arrive at preliminary dimensions is illustrated below by means of an example drawn from our consultancy practice.

*Example 3.1*

The following data relate to the Falaknaaz Shopping Centre at Dubai (Fig. 3.3):

$l_x = 24$ m, $l_y = 27$ m
Structural depth, $h = 1.07$ m
Grid dimension 1.50 m
Topology: square over square offset
Design loading 1 kN/m$^2$
Load at ultimate limit, $p = 1.50$ kN/m$^2$

The space frame is supported on the top nodes along all four sides. We can estimate the maximum bending moment and deflection and select preliminary tube sizes for the top and bottom chords and diagonal bracing using the method of limit states and Eurocode 3 (Curve b).

*Solution.* The yield strength of the steel tubes is 240 MPa and the aspect ratio is

$$\alpha = \frac{l_y}{l_x} = \frac{27}{24} = 1.125$$

From Table 3.1, the interpolated values of $k_w$ and $k_M$ are 99.85 and 10.575, respectively.

The maximum compression will govern the selection of the chord members:

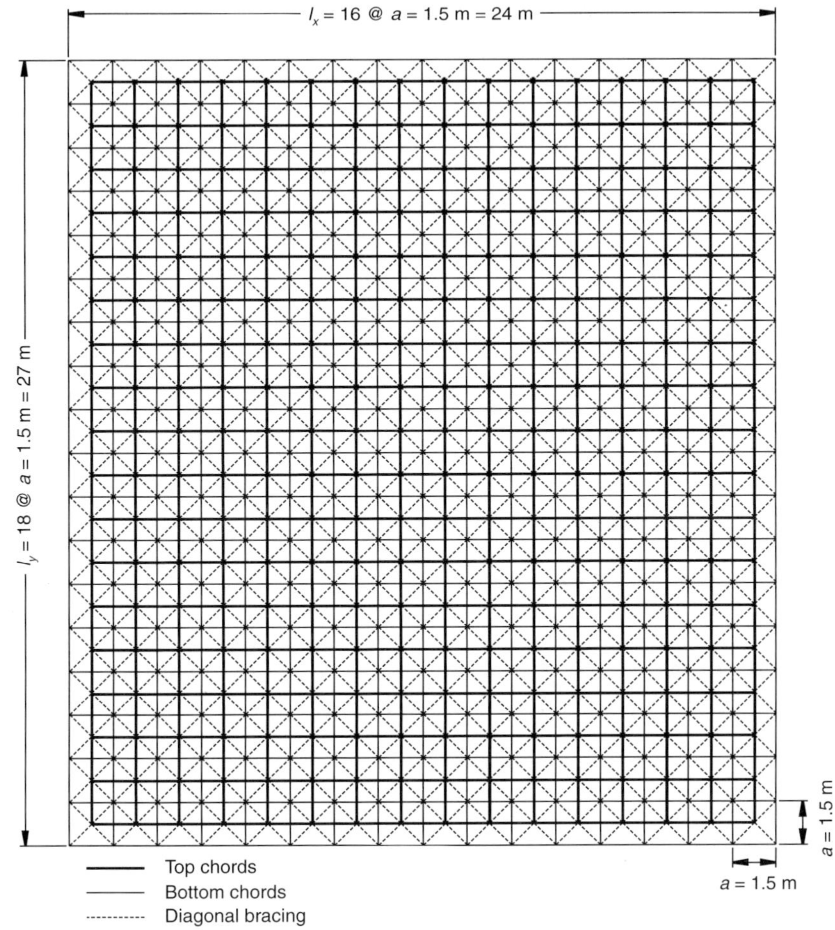


*Fig. 3.3. The space frame for the Falaknaaz Shopping Centre, Dubai*

$$\frac{1.50 \times 24^2 \times 1.50}{10.575 \times 1.07} = 115 \text{ kN}$$

Let us try a tube of 76.20 mm × 2.90 mm, with area $A = 667$ mm$^2$ and a radius of gyration $r = 25.92$ mm. Then the rigidity of the equivalent plate is given by

$$D = \frac{AE}{a}\frac{h^2}{2} = \frac{667 \times 210 \times 1.07^2}{1.5 \times 2}$$
$$= 53\,455$$

The estimated maximum deflection is

$$\frac{pl_x^4}{k_w D} = \frac{1.50 \times 24^4 \times 1000}{99.5 \times 53\,455} = 93.56 \text{ mm}$$

We now use Eurocode 3, Curve b, to compute the permissible compression that a tube of size 76.20 mm × 2.90 mm can carry at the ultimate limit state:

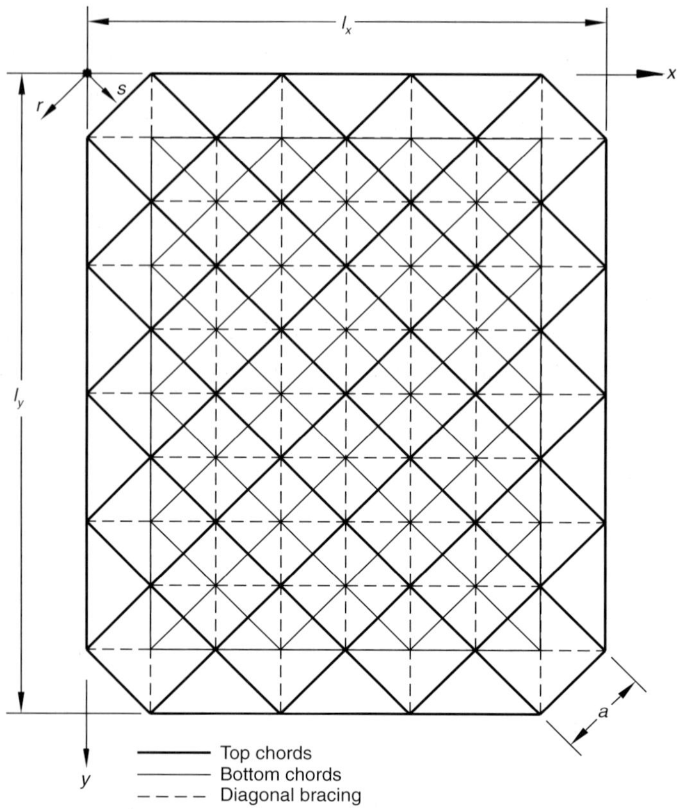


*Fig. 3.4. Square over square diagonally offset topology*

$$\lambda_e = \sqrt{\frac{210\,000}{240}} = 92.93$$

The equivalent length $l_e$ is $0.85 \times 1500 = 1275$ mm, and

$$\lambda = \frac{l_e}{r} = \frac{1275}{25.92} = 49.19$$

$$\frac{\lambda}{\lambda_e} = \lambda_{rel} = 0.53$$

From Table 2.1 (Eurocode 3, Table b), $\omega_{buc} = 0.8701$, and thus the allowable compression at the ultimate limit state, is

$$\frac{0.8701 \times 667 \times 240}{1000} = 139 \text{ kN} > 115 \text{ kN}$$

Hence for the first run of the structural analysis we adopt a tube size of 76.20 mm × 2.90 mm for the top and bottom chords. There is no simple rule for arriving at the size of diagonal bracing. From experience we adopt a size of 50.80 mm × 2.90 mm.

*Note.* The structural analysis was carried out assuming these preliminary sizes and the design was optimized by successive iteration. The maximum compressive force in the optimized design was found to be 117.72 kN, which is in close agreement with the value of 115 kN estimated by means of the plate analogy.

The maximum deflection of the optimized space frame was 127.70 mm, which is higher than the estimated deflection of 93.56 mm. This higher value is due to the fact that in the preliminary run we assumed that all top and bottom chords were 76.20 mm × 2.90 mm in size. In the final design, all except 31 members out of the total of 564 turned out to be 50.80 mm × 2.90 mm in size. Hence it will be more appropriate to use the plate rigidity of 50.80 mm × 2.90 mm when estimating the expected deflection by using the plate analogy. Noting that the area of the tube is 50.80 × 2.90 = 436 mm$^2$,

$$D = \frac{436 \times 210 \times 1.06^2}{2 \times 1.5}$$
$$= 34\,292$$

Hence

$$w_{\max} = \frac{1.50 \times 24^2 \times 1000}{99.5 \times 34\,292}$$
$$= 145.85 \text{ mm}$$

## 3.4. Plate analogy for a double-layer flat space frame with square over square diagonally offset topology

Referring to Fig. 3.4, the governing partial differential equation of the equivalent plate may be derived [3.6] as

$$\frac{\partial^4 w}{\partial x^4} + 4\frac{\partial^4 w}{\partial x^2\,\partial y^2} + \frac{\partial^4 w}{\partial y^4} = \frac{p}{D} \tag{3.8}$$

where

$$D = \frac{EA}{2a}\frac{h^2}{2}$$

The solution for $w$ is

$$w = \frac{16 p l_x^4}{\pi^6 D} \sum_{m=1,3,5,\ldots}^{\infty} \sum_{n=1,3,5,\ldots}^{\infty} \frac{\sin m\pi\xi \sin n\pi\eta}{mn(m^4 + 4m^2n^2/\alpha^2 + n^4/\alpha^4)} \tag{3.9}$$

where $\xi = x/l_x$ and $\eta = y/l_y$. The maximum vertical deflection occurring at $x = l_x/2$ and $y = l_y/2$ may be written as

$$w_{\max} = \frac{16 p l_x^4}{\pi^6 D} \sum_{m=1,3,5,\ldots}^{\infty} \sum_{n=1,3,5,\ldots}^{\infty} \frac{-1^{(m+n)/2-1}}{mn(m^4 + 4m^2n^2/\alpha^2 + n^4/\alpha^4)} \tag{3.10}$$

The expression for $w_{max}$ may be rewritten as

$$w_{max} = \frac{pl_x^4}{k_w D} \tag{3.11}$$

In what follows, the term

$$mn(m^4 + 4m^2n^2/\alpha^2 + n^4/\alpha^4)$$

will be denoted by $N$.

The values of $k_w$ can be tabulated for different values of $\alpha$ by summing up the series. Expressions for the bending moments $M_x$, $M_y$ and $M_{xy}$ are found as follows:

$$M_x = D\frac{\partial^2 w}{\partial x^2} = n_x h = \frac{16pl_x^2}{\pi^4}\sum_m\sum_n \frac{1}{N} m^2 \sin m\pi\xi \sin n\pi\eta \tag{3.12}$$

$$M_y = D\frac{\partial^2 w}{\partial y^2} = n_y h = \frac{16pl_x^2}{\pi^4}\sum_m\sum_n \frac{1}{N}\frac{n^2}{\alpha^2} \sin m\pi\xi \sin n\pi\eta \tag{3.13}$$

$$M_{xy} = D\frac{\partial^2 w}{\partial x\,\partial y} = n_{xy} h = \frac{16pl_x^2}{\pi^4}\sum_m\sum_n \frac{1}{N}\frac{mn}{\alpha} \cos m\pi\xi \cos n\pi\eta \tag{3.14}$$

By forming the equation of equilibrium of the element shown in Fig. 3.5 in the direction of the force $P_r$, it can be seen that

$$P_r = n_x \frac{2a}{\sqrt{2}}\frac{1}{\sqrt{2}} + n_y \frac{2a}{\sqrt{2}}\frac{1}{\sqrt{2}} - 2n_{xy}\frac{2a}{\sqrt{2}}\frac{1}{\sqrt{2}}$$

or

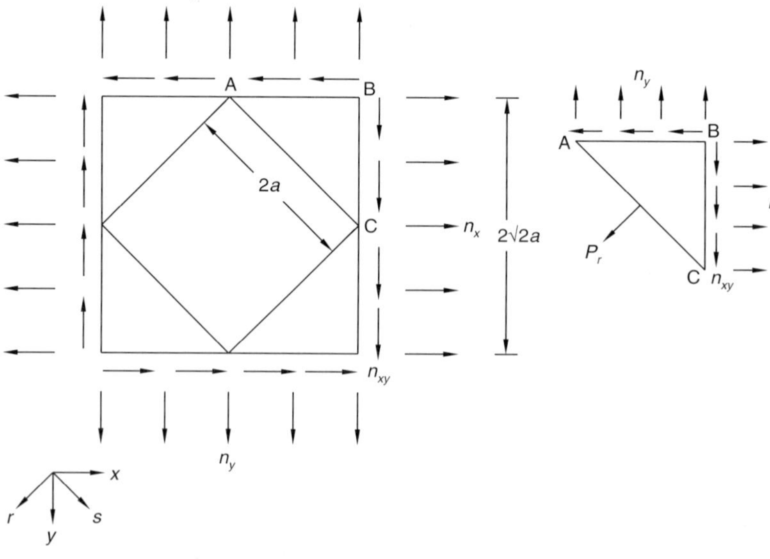


*Fig. 3.5. The components of the force* $P_r$

*Table 3.2. Values of $\alpha$, $k_w$, $k_M$ and $k_E$*[a]

| | | | | | | | |
|---|---|---|---|---|---|---|---|
| $\alpha$ | 1.0 | 1.2 | 1.4 | 1.6 | 1.8 | 2.0 | 3.0 |
| $k_w$ | 372 | 258 | 204 | 168 | 140 | 124 | 87 |
| $k_M$ | 19.4 | 17.6 | 15.8 | 14.1 | 12.5 | 11.5 | 9.3 |
| $k_E$ | 18.9 | 15.7 | 14.5 | 13.6 | 12.8 | 12.6 | 13.2 |

[a]Reproduced with permission from White and Gladischefski [3.6]

$$P_r = 2a\left(\frac{n_x + n_y}{2} - n_{xy}\right) \tag{3.15}$$

Similarly,

$$P_s = 2a\left(\frac{n_x + n_y}{2} + n_{xy}\right) \tag{3.16}$$

Hence

$$P_r = \frac{16pl_x^2}{\pi^4}\frac{a}{h}\sum_m\sum_n\frac{1}{N}\times \left[\left(m^2 + \frac{n^2}{\alpha^2}\right)\sin m\pi\xi \sin n\pi\eta - \frac{2mn}{\alpha}\cos m\pi\xi \cos n\pi\eta\right] \tag{3.17}$$

$$P_s = \frac{16pl_x^2}{\pi^4}\frac{a}{h}\sum_m\sum_n\frac{1}{N}\times \left[\left(m^2 + \frac{n^2}{\alpha^2}\right)\sin m\pi\xi \sin n\pi\eta + \frac{2mn}{\alpha}\cos m\pi\xi \cos n\pi\eta\right] \tag{3.18}$$

The expression for maximum $P_r$ may be written as

$$\max P_r = \mp\frac{pl_x^2}{k_M}\frac{a}{h} \tag{3.19}$$

The minus sign denoting compression applies to the top chords and the plus sign denoting tension refers to the bottom chords. The subscript M in $k_M$ refers to the middle. Apart from the moments at the centre, one would also be interested in the moments that develop at the ends. Denoting these by $P_E$, we can write

$$\max P_E = \pm\frac{pl_x^2}{k_E} \tag{3.20}$$

where the subscript E in $k_E$ refers to the ends.

The coefficients $k_w$, $k_M$ and $k_E$ for different values of $\alpha$ are presented in Table 3.2.

### *Example 3.2*

The following data relate to alternative design No. 1 proposed for the Al Wahda Sports Club at Abu Dhabi, United Arab Emirates (Fig. 3.6):

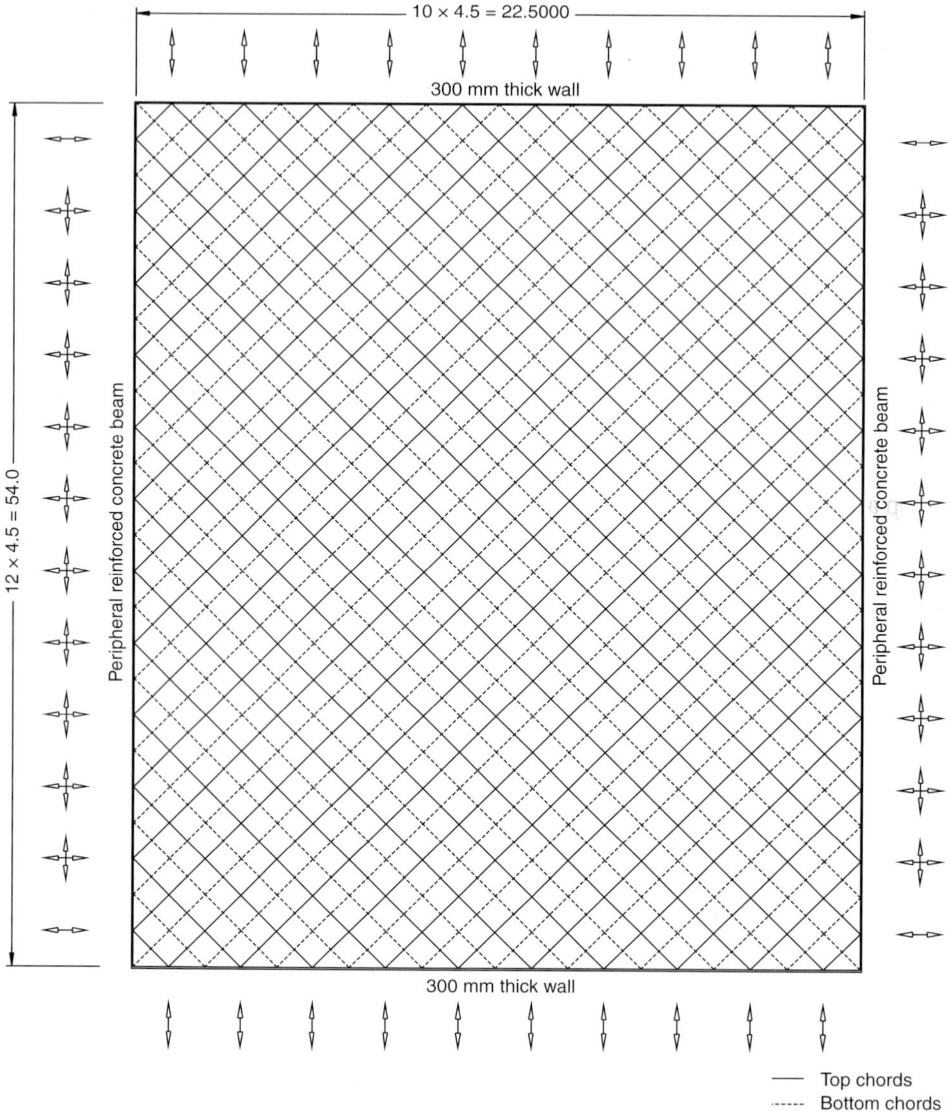


*Fig. 3.6. The space frame for the Al Wahda Sports Club, Abu Dhabi*

Frame type: flat double-layered
$l_x = 45$ m, $l_y = 54$ m
Structural depth 2.25 m
Load at ultimate limit state 2.25 kN/m$^2$
Grid dimension, $a = 3.18$ m
Topology: square over square diagonally offset
Yield strength of tubes 450 N/mm$^2$
Support conditions: supported on all four sides

Our task is to calculate the size of the tubes for the top and bottom chords, for the preliminary analysis, and to estimate the maximum deflection.

*Solution*.

$$\alpha = \frac{54}{45} = 1.2$$

From Table 3.2, the value of $k_M$ is 17.6, and hence the maximum compression $F_{max}$ in the top chords is given by

$$F_{max} = \frac{2.25 \times 45^2}{17.6} \frac{3.18}{2.25} = 366 \text{ kN}$$

Using the EURO B program on the accompanying CD-ROM, it is found that a tube of 114.30 mm × 4.50 mm will have an allowable compression of 406 kN, which is higher than the required value of 366 kN. Thus this tube size will be adopted both for the top and bottom chords. From past experience, a size of 76.10 mm × 3.25 mm is chosen for the bracing diagonals. The first cycle of analysis is carried out assuming these preliminary sizes.

We can now estimate the deflection using the following data:

Area of tube, $A = \pi(114.30 - 4.50)4.50 = 1552 \text{ mm}^2$
$E = 210 \text{ kN/mm}^2$
$h = 2.25$ m

The flexural rigidity of the equivalent plate is given by

$$D = \frac{AE}{2a} \frac{h^2}{2}$$

$$= \frac{1552 \times 210 \times 2.25^2}{3.18 \times 2 \times 2} = 129\,715$$

Referring to Table 3.2, for $\alpha = 1.2$ the value of $k_w$ is 258. Thus the estimated deflection is

$$\frac{2.25 \times 45^4 \times 1000}{258 \times 129\,715} = 276 \text{ mm}$$

When this space frame is optimized to minimum weight, starting with these preliminary sizes, it has a maximum deflection of 470 mm. This is because the preliminary design is very stiff and all top and bottom chords are of the same chosen size to carry the maximum compression. The structure softens progressively as the optimization process proceeds to arrive at the minimum weight.

## 3.5. Double-layer space frame with diagonal over square topology

No published analytical solutions or tables are available to arrive at maximum forces and deflections. An approximate approach to obtain preliminary values for the forces is illustrated in the example given below.

*Example 3.3*

A double-layer space frame, 16.8 m × 16.8 m in size, with a diagonal over square topology is supported at the top nodes on columns at 8.4 m centres on all four sides. It carries a downward load of 1.5 kN/m$^2$. Our task is to calculate the preliminary sizes of the tubes for the top and bottom chords. The yield strength of tubes is 310 MPa, the structural depth is 1.4 m and the grid dimension $a$ is 2.8 m for the bottom chords.

*Solution*. Consider a strip of the space frame equal to the grid dimension of 2.8 m and regard it as a beam (Fig. 3.7). The central bending moment is

$$\frac{1.5 \times 2.8 \times 16.8^2}{8} = 148.20 \text{ kN-m}$$

This bending moment is resisted by the couple formed by $C$ and $T$, the total compression and total tension, respectively. The tension $T$ in the bottom chords is 148.20/1.4 = 105.85 kN.

The total tension is equivalent to the total compression:

$$T = \frac{2C}{\sqrt{2}} = \sqrt{2}C$$

The compression $C$ in the top chords is 74.80 kN, and thus compression will govern the design. The length of the top chord is 2.8/√2 = 1.98 m = 1980 mm.

Let us try a tube of 60.30 mm × 3.25 mm. The allowable compression computed using the EURO B formula is 106 kN, which is greater than the required value of 74.80 kN.

Hence a suitable tube is 60.3 mm × 3.25 mm with an area

$$\pi(60.30 - 3.25)3.25 = 582 \text{ mm}^2$$

which gives an allowable tension of

$$\frac{582 \times 310}{1000} = 180.42 \text{ kN}$$

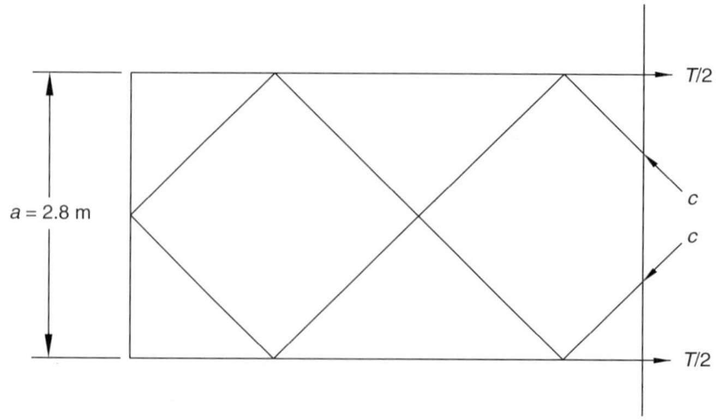


*Fig. 3.7. Diagonal over square topology*

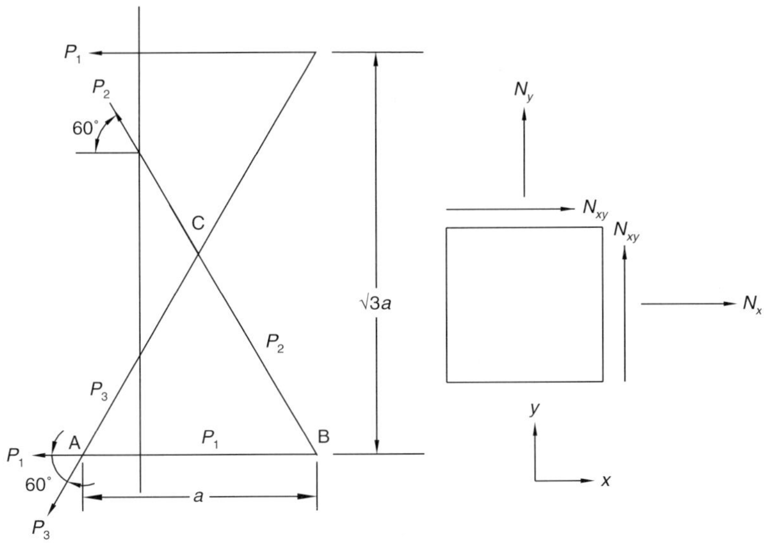


*Fig. 3.8. The tensile forces* $P_1$, $P_2$ *and* $P_3$ *in the members of an equilateral grid and the equivalent membrane with the resultant in-plane stresses* $N_x$, $N_y$ *and* $N_{xy}$

Our selected tube size is thus suitable. The same size will be used for both the top and the bottom chords in the first cycle of the analysis. For the bracings one may adopt a size of 42.40 mm × 3.25 mm.

## 3.6. Triangular three-way grids

Triangular three-way grids are often more economical than two-way grids for flat space frames. They are also the preferred configuration for braced barrel vaults and domes. The configuration normally chosen is equilateral. In view of their importance, we shall proceed to derive the continuum equivalents of flat space frames, reticulated barrel vaults and domes involving three-way grids composed of equilateral triangles. The derivations that follow are based on the landmark paper by Wright [3.7] which is remarkable for its simplicity and structural intuition.

### 3.6.1. *Double layer flat space frame with equilateral triangle over equilateral triangle offset topology*

Figure 3.8 shows the tensile forces $P_1$, $P_2$ and $P_3$ in the members AB, BC and CA of an equilateral grid with sides of length $a$, and the equivalent membrane with the resultant in-plane stresses $N_x$, $N_y$ and $N_{xy}$ acting on it. The equilateral grid is cut by a vertical plane of length $a$.

Resolving the forces in the $x$ direction,

$$2P_1 + \frac{P_2}{2} + \frac{P_3}{2} = \sqrt{3}aN_x \tag{3.21a}$$

or

$$N_x = \frac{4P_1 + P_2 + P_3}{2\sqrt{3}a} \tag{3.21b}$$

Similarly,

$$N_y = \frac{(P_2 + P_3)\sqrt{3}}{2a}$$

$$N_{xy}\sqrt{3}a = P_2 - P_3\frac{\sqrt{3}}{2} \tag{3.21c}$$

or

$$N_{xy} = \frac{P_2 - P_3}{2a} \tag{3.21d}$$

Solving for $P_1$, $P_2$ and $P_3$ from equations (3.21a) to (3.21d) we obtain

$$P_1 = \frac{a}{2\sqrt{3}}(3N_x - N_y) \tag{3.22a}$$

$$P_2 = \frac{a}{\sqrt{3}}(N_y + \sqrt{3}N_{xy}) \tag{3.22b}$$

$$P_3 = \frac{a}{\sqrt{3}}(N_y - \sqrt{3}N_{xy}) \tag{3.22c}$$

Following Wright [3.7], the material of the analogous membrane is assumed to be anisotropic to start with. Members AB, BC and CA have the same area of cross-section $A$ and modulus of elasticity $E$. Keeping the point A and the direction AB fixed, it is possible to plot the new positions of the vertices A, B and C after the grid deforms under the forces $P_1$, $P_2$ and $P_3$. The new positions $a$, $b$ and $c$ of the vertices A, B and C are found by constructing a Williot–Mohr diagram, as shown in Fig. 3.9. All the displacements shown in the diagram must be multiplied by $a/AE$ to arrive at the actual displacements.

The expression for $cN$ is

$$JH = P_2 - P_1\cos 60 - P_3\cos 60$$

$$= P_2 - \frac{P_1}{2} - \frac{P_3}{2}$$

$$\frac{JH}{Jc} = \cos 30 = \sqrt{3}/2$$

Hence,

$$Jc = \frac{1}{\sqrt{3}}(2P_2 - P_1 - P_3)$$

$$cM = Jc\sin 30 = \frac{1}{2\sqrt{3}}(2P_2 - P_1 - P_3)$$

$$MN = P_3\frac{\sqrt{3}}{2}$$

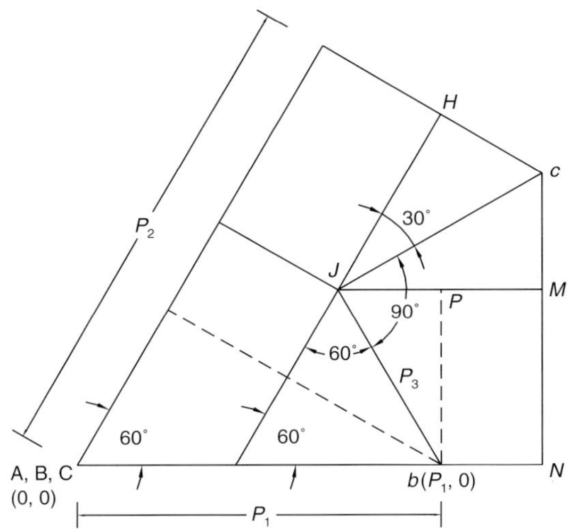


*Fig. 3.9. Williot–Mohr diagram showing the new positions a, b and c of the vertices A, B and C*

Hence

$$cN = cM + MN = \frac{1}{2\sqrt{3}}(2P_2 - P_1 + 2P_3) \tag{3.23}$$

The expression for $aN$ is

$$\begin{aligned} aN &= ab + bN = P_1 + JM - JP \\ &= P_1 + Jc\cos 30 - P_3 \sin 30 \\ &= P_1 + \frac{1}{2}(2P_2 - P_1 - 2P_3) \\ &= \frac{1}{2}(P_1 + 2P_2 - 2P_3) \end{aligned} \tag{3.24}$$

To summarize, B has moved to the right in the $x$ direction by $P_1a/AE$. It has not moved in the $y$ direction. The vertex $c$ has moved by an amount $aN/AE$ (i.e. by $(a/2AE)(P_1 + 2P_2 - 2P_3)$ in the positive $x$ direction) and by $cN/AE$ (i.e. by $(-P_1 + 2P_2 + 2P_3)(a/2\sqrt{3}AE)$ in the positive $y$ direction). These movements have resulted in the following strains:

$$\varepsilon_x = \frac{P_1}{AE} \tag{3.25a}$$

and

$$\varepsilon_y = \frac{-P_1 + 2P_2 + 2P_3}{3AE} \tag{3.25b}$$

The angular distortion causing shear strain is given by the differential displacement of $c$ and $b$ in the horizontal direction divided by the vertical distance between $b$ and $c$. The differential horizontal displacement is $(P_2 - P_3)a/AE$, and the vertical distance is $(\sqrt{3}/2)a$. Hence,

$$r_{xy} = \frac{2(P_2 - P_3)}{\sqrt{3}AE} \tag{3.25c}$$

The expression for the strain in terms of the stresses may be written as

$$\varepsilon_x = \frac{\sigma_x}{E'_x} - \nu'_x \frac{\sigma_y}{E'_y} \tag{3.26a}$$

$$\varepsilon_y = \frac{\sigma_y}{E'_y} - \nu'_y \frac{\sigma_x}{E'_x} \tag{3.26b}$$

$$r_{xy} = \frac{\tau_{xy}}{G'} \tag{3.26c}$$

or

$$\sigma_x = \frac{E'_x}{1 - \nu'_x \nu'_y}(\varepsilon_x + \nu'_x \varepsilon_y) \tag{3.27a}$$

$$\sigma_y = \frac{E'_y}{1 - \nu'_x \nu'_y}(\varepsilon_y + \nu'_y \varepsilon_x) \tag{3.27b}$$

$$\tau_{xy} = G' r_{xy} \tag{3.27c}$$

In the above expressions the primed quantities relate to the equivalent continuum, $E$ is Young's modulus, $G$ is the modulus of rigidity and $\nu$ is the Poisson ratio of the space frame material. The elastic constants appearing in equations (3.26) and (3.28) may be determined from equations (3.21), (3.25) and (3.26) by noting that

$$\sigma_x = \frac{N_x}{t'} \tag{3.28a}$$

$$\sigma_y = \frac{N_y}{t'} \tag{3.28b}$$

$$\tau_{xy} = \frac{N_{xy}}{t'} \tag{3.28c}$$

where $t'$ is the thickness of the analogous membrane or lamina.

From equations (3.28a) to (3.28c) and (3.21),

$$\sigma_x = \frac{4P_1 + P_2 + P_3}{2\sqrt{3}at'} \tag{3.29}$$

and when $\sigma_y = 0$ and $P_2 + P_3 = 0$ we have

$$\sigma_x = \frac{2P_1}{\sqrt{3}at'}$$

With $\sigma_y = 0$, and using equation (3.26a),

$$E'_x = \frac{\sigma_x}{\varepsilon_x}$$

Noting that $\varepsilon_x = P_1/AE$, from equation (3.25a)

$$E'_x = \frac{2P_1}{\sqrt{3}at'}\frac{AE}{P_1} = \frac{2AE}{\sqrt{3}at'} \tag{3.30}$$

Similarly, $E_y'$ may be determined as $\sigma_y/\varepsilon_x$ for $\sigma_x = 0$, and the result is

$$E'_y = \frac{2AE}{\sqrt{3}at'} \tag{3.31}$$

Hence, $E_x' = E_y' = E'$, where $E'$ is the modulus of elasticity of the equivalent continuum plate or shell.

From equation (3.21d),

$$\tau_{xy} = \frac{N_{xy}}{t'} = \frac{P_2}{at'}$$

with $\sigma_y = 0$ and $P_2 + P_3 = 0$. Similarly, from equation (3.25c),

$$r_{xy} = \frac{4P_2}{\sqrt{3}AE}$$

under the same conditions. Hence

$$G' = \frac{\tau_{xy}}{r_{xy}} = \frac{\sqrt{3}AE}{4at'} \tag{3.32}$$

setting $\varepsilon_x = 0$ and $P_1 = 0$. Also, from equation (3.26a),

$$\nu'_x = \frac{\sigma_x}{E'_x}\frac{E'_y}{\sigma_y}$$

Noting that $E_x' = E_y'$ and

$$\frac{\sigma_x}{\sigma_y} = \frac{1}{2\sqrt{3}a}\frac{2a}{\sqrt{3}}$$

then

$$\nu'_x = \frac{1}{3} \tag{3.33}$$

By similar reasoning, it can be shown that

$$\nu'_y = \frac{1}{3}$$

We have thus demonstrated that the material of the analogous plate or shell is isotropic, with a Poisson ratio of 1/3.

### *3.6.2. The flexural rigidity of an analogous plate for equilateral triangular grids*

Consider an analogous plate of unit width subject to a bending moment $M$. Let the flexural rigidity be $D$ (Fig. 3.10). The strain in the top lamina is

$$\frac{(R+h/2)\theta - R\theta}{R\theta} = \frac{h}{2R} \tag{3.34}$$

The force per unit width is given by $(h/2R)E't$. The bending moment is

$$M = \frac{h^2}{2R}\frac{E'}{1-\nu^2}t' = \frac{D}{R}$$

and

$$\nu = \frac{1}{3}$$

Hence,

$$D = \frac{h^2}{2}\frac{9}{8}E't'$$

Substituting for $E't'$ from equation (3.30),

$$D = \frac{h^2}{2}\frac{9}{8}\frac{2AE}{\sqrt{3}a}$$
$$= \frac{9AE}{4\sqrt{3}a}\frac{h^2}{2} \tag{3.35}$$

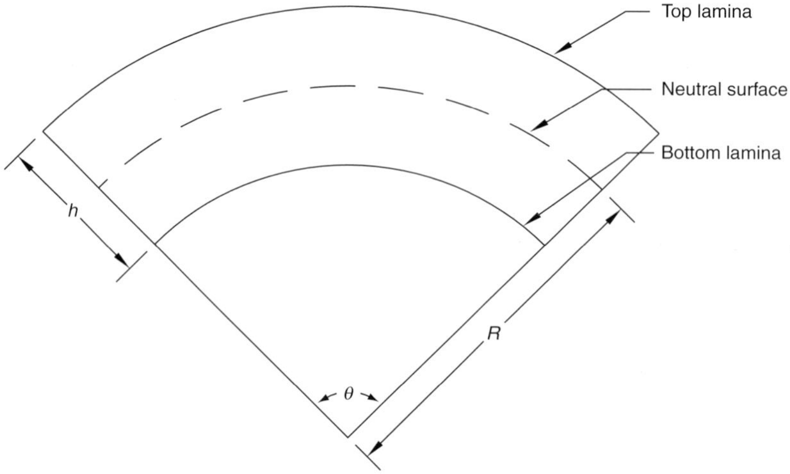


*Fig. 3.10. Analogous plate of unit width*

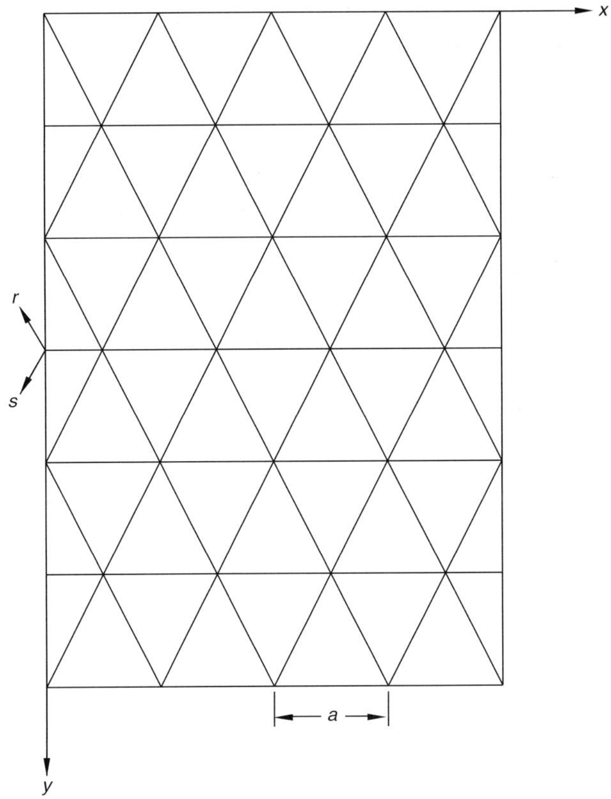


*Fig. 3.11. Triangular three-way grid*

*Table 3.3. Values of* $\alpha$, $k_w$, $k_x$, $k_R$ *and* $k_E$[a]

| $\alpha$ | 1.0 | 1.2 | 1.4 | 1.6 | 1.8 | 2.0 |
|---|---|---|---|---|---|---|
| $k_w$ | 246 | 177 | 142 | 120 | 107 | 98 |
| $k_x$ | 35.2 | 25.3 | 19.7 | 16.6 | 14.7 | 13.7 |
| $k_R$ | 35.2 | 33.3 | 32.8 | 33.3 | 34.2 | 35.0 |
| $k_E$ | 33.3 | 27.8 | 29.5 | 25.1 | 23.8 | 22.7 |

[a]Reproduced with permission from White and Gladischefski [3.6]

The preliminary design of three-way grids with an equilateral triangle over equilateral triangle offset topology (Fig. 3.11) can be carried out using the data given in Table 3.3:

$$\text{Maximum } w = \frac{pl_x^4}{Dk_W}$$

where

$$D = \frac{9AE}{4\sqrt{3}a}\frac{h^2}{2}$$

In Table 3.3, $k_w$ is tabulated for different values of $\alpha$.

The maximum forces in the $r$ direction in the grid members at mid-span are given by

$$F_r = \mp \frac{pl_x^2}{k_R} \frac{a}{h} \tag{3.36}$$

The minus sign denotes compression in the top chords and the plus sign denotes tension in the bottom chords. Similarly, the maximum force at mid-span in the $x$ direction is given by

$$F_x = \mp \frac{pl_x^2}{k_x} \tag{3.37}$$

The maximum forces at the ends in the $r$ direction are given by

$$F_r = \pm \frac{pl_x^2}{k_E}$$

### 3.6.3. Effective thickness and Young's modulus of analogous shells

It can be easily shown that analogous plates or shells corresponding to equilateral triangular grids exhibit orthotropic behaviour, not only with respect to in-plane forces but also to other general deformations, including bending.

Let us consider bending in the $x$ direction (Fig. 3.12). A curvature of $1/R_x$ caused by bending causes a curvature of $1/4R_x$ in elements 2 and 3. The curvature of $1/R_x$ causes bending of $EI/R_x$ in element 1. In elements 2 and 3, resisting moments of $\frac{1}{4}(EI/R_x)$ are induced. The moment vectors acting on a cut of $\sqrt{3}(a)$ are shown in Fig. 3.12. Resolving them in the $x$ direction, the total bending moment in the $x$ direction is $(9/4)(EI/R_x)$. Hence the bending moment per unit width $m$ may be found by dividing by $\sqrt{3}(a)$ to give

$$m_x = \frac{9}{4} \frac{EI}{R_x} \frac{1}{\sqrt{3}a} = \frac{3\sqrt{3}EI}{4R_x a} \tag{3.38}$$

By similar reasoning it can be shown that if a curvature $1/R_y$ is applied in the $y$ direction the resulting bending moment is given by

$$m_y = \frac{3\sqrt{3}EI}{4R_y a} \tag{3.39}$$

demonstrating isotropic behaviour under bending deformation.

Let us consider the analogous plate or shell corresponding to an equilateral triangular grid submitted to the action of a moment $m$ in one direction:

$$m = \frac{D}{R} = \frac{E't'^3}{12R(1-\nu^2)}$$

or

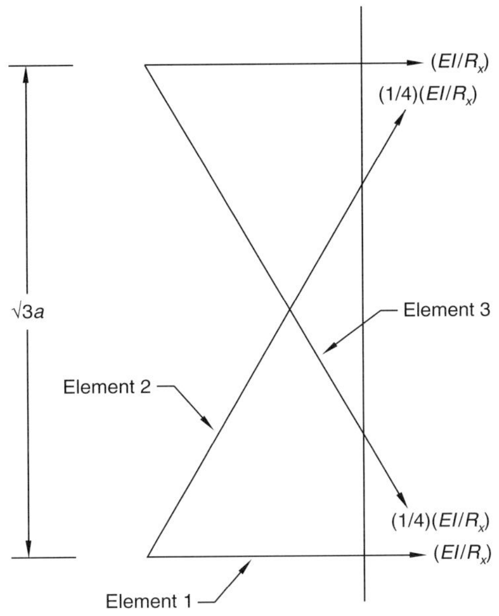


*Fig. 3.12. Bending in the* x *direction in an analogous plate or shell corresponding to an equilateral triangular grid*

$$t'^3 = \frac{12mR(1-\nu^2)}{E'} \tag{3.40}$$

Noting that $\nu = 1/3$, and substituting for $m$ from equation (3.38) and for $E'$ from equation (3.30),

$$t' = 2\sqrt{3}\sqrt{\frac{I}{A}} \tag{3.41}$$

Noting that $I = Ar^2$, where $r$ is the radius of gyration,

$$t' = 2\sqrt{3}r \tag{3.42}$$

Having found $t'$ explicitly, it is possible to derive the modulus of elasticity of the analogous plate or shell from equation (3.30). Hence,

$$E' = \frac{2AE}{\sqrt{3}a}\frac{1}{2\sqrt{3}r} = \frac{AE}{3ra} \tag{3.43}$$

Let us suppose that an equilateral grid of 3000 mm is composed of tubes of outside diameter $D = 76.20$ mm and thickness $t = 3.25$ mm. Then the area is given by

$$A = \pi(D - t)t$$

and

$$r = \frac{D-t}{2\sqrt{2}}$$

Hence,

$$E' = \frac{\pi(D-t)t}{3\times 3000}\frac{2\sqrt{2}E}{D-t} = \frac{\pi\times 3.25\times 2\sqrt{2}\times E}{3\times 3000} = 0.0032E$$

The equivalent thickness is then

$$2\sqrt{3}\frac{D-t}{2\sqrt{2}} = 89 \text{ mm}$$

or the thickness of the analogous plate or shell is higher than the diameter of the tubes forming the reticulated structure. On the other hand, the modulus of elasticity of the material of the analogous shell is only a small fraction of the modulus of elasticity of the members comprising the equilateral grid. The situation was aptly summarized by Wright [3.7] who concluded that an analogous plate or shell functions as a fairly thick structure made of a spongy material.

*Example 3.4*

A three-way double-layer flat space frame roof of 44 m × 41.57 m carries a load of 1.5 kN/m$^2$ at the ultimate limit state and is supported as shown in Fig. 3.13. The triangular grids formed by the top and bottom chords are equilateral. The grid dimension is 4 m and the structural depth is 3.266 m. The yield strength of the tubes is 240 MPa. Our aim is to determine the preliminary sizes of the top and bottom chords and to estimate the maximum deflection.

*Solution*. The aspect ratio is 44/41.57 = 1.05 ≈ 1. Thus maximum compression will govern. From Table 3.3

$$F_{\max} = \frac{1.5\times 41.57^2\times 4000}{33\times 3266} = 96 \text{ kN}$$

Let us try a tube of dimensions 88.90 mm × 4.05 mm. The safe compression that it can carry according to Eurocode 3, Curve b, is 121.27 kN. Hence a tube of these dimensions is safe.

The estimated deflection for an area of 1080 mm$^2$ is computed as follows:

$$D = \frac{9AE}{4\sqrt{3}a}\frac{h^2}{2}$$

$$= \frac{9\times 210\times 1080}{4\sqrt{3}\times 4}\frac{3.266^2}{2} = 392\,832$$

Referring to Table 3.3, $k_w = 246$, and thus the estimated deflection is

$$\frac{1.50\times 41.57^4\times 1000}{246\times 392\,832} = 46.35 \text{ mm}$$

The maximum compression of 88.64 kN found by computer analysis is in close agreement with the estimated maximum compression of 96 kN. The maximum deflection found by actual analysis is 62 mm.

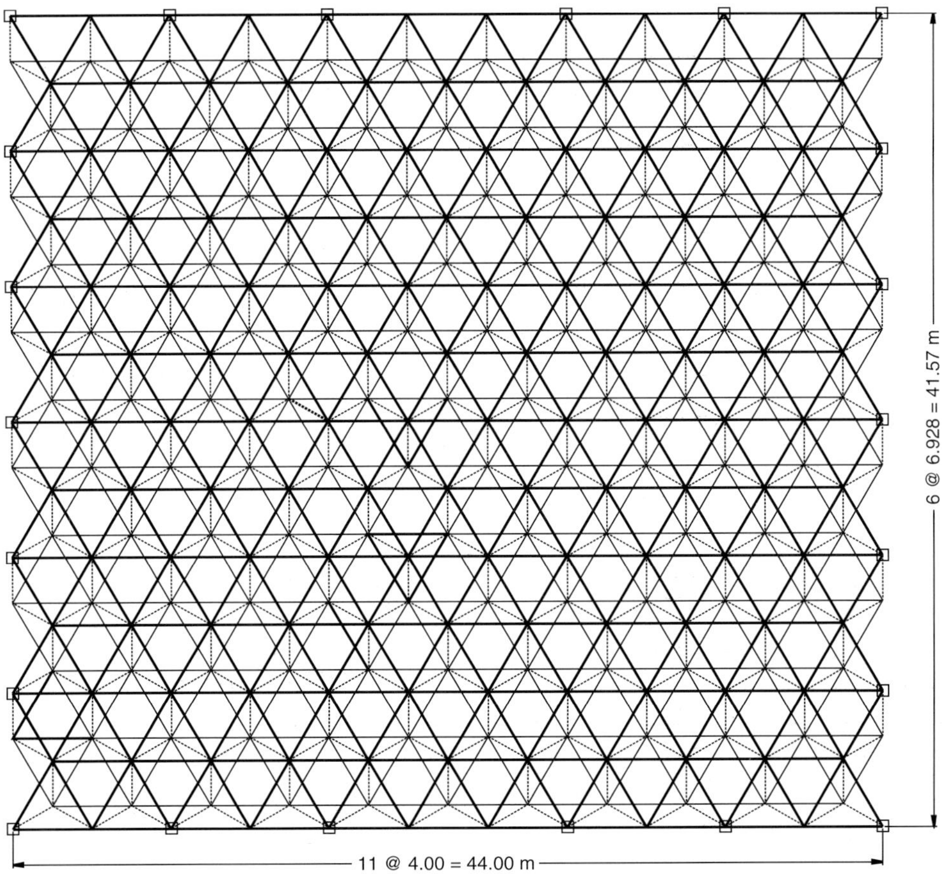


*Fig. 3.13. The support of the three-way double-layer flat space frame roof in Example 3.4*

## 3.7. References


3.1. KOLLAR, L. Analysis of double-layer space trusses by the equivalent continuum method. *Proceedings of the 2nd International Conference on Space Structures*, University of Surrey, Guildford, 1975, pp. 73–76.

3.2. HEKI, K. and SAKA, T. Stress analysis of lattice plates as anisotropic continuum plates. *Proceedings of the 1971 IASS Symposium on Tension Structures and Space Frames*, Architectural Institute of Japan, Tokyo, 1972, pp. 663–674.

3.3. RENTON, J. D. The related behaviour of plane grids, space grids and plates. In: R. M. Davies (ed.) *Space Structures: International Conference on Space Structures*, University of Surrey, Blackwell Scientific, Oxford, 1967, pp. 19–32.

3.4. RENTION, R. D. General properties of space grids *International Journal of Mechanical Science*, 1970, **12**, 801–810.

3.5. SOARE, M. V. Contributions to the analysis of double-layer mesh grids. *2nd International Conference on Space Structures*, University of Surrey, Guildford, 1975, pp. 159–168.

3.6. WHITE, H. and GLADISCHEFSKI, H. Einfache Regein Zur Vorbemessung Von Raumfachwerken [Approximate design of space structures]. *Stahl*, 1981, **1**, 110.

3.7. WRIGHT, D. T. Membrane forces and buckling in reticulated shells. *Journal of the Structural Division, Proceedings of the ASCE*, 1965, **ST1**, 173–201.

# 4. Double-layer flat space frames

## 4.1. Introduction

Double-layer flat space frames may consist of two-way or three-way grids. In a *two-way grid*, members running in two orthogonal directions intersect at a node. In *three-way grids*, which are normally equilateral, members running in three directions will intersect at a node. Six configurations that are normally used for two-way double layer space frames are shown in Fig. 2.8 of Chapter 2. The optimal topology for a given situation will be governed by the spans, loads and support conditions. However, a few general remarks on the relative merits of these arrangements may be in order.

Two-way grids with dense configurations such as square on square offset (type 1) and square on square offset, set diagonally (type 2), possess higher rigidity than sparse configurations such as square on diagonal (type 5), diagonal on square (type 6) and square on larger square (type 3). The square on square offset, set diagonally (type 2), is more rigid than the square on square offset (type 1). Configuration types 1 and 2 may often be more rigid than required.

To produce competitive designs, it is imperative to choose the most appropriate topology that, while reducing the consumption of steel to a minimum, is rigid enough to control deflections within prescribed limits. An example will best illustrate this point. For the Al Wahda Sports Hall space frame roof, which is dealt with in detail in the case study that follows, we were obliged to prune consumption of steel to a minimum because we had to win the job against stiff international competition. With this end in view, a square on diagonal configuration that offered just the right amount of rigidity and no more was selected.

Where very high rigidity is called for because of large spans, heavy loading or infrequent supports, three-way grids such as the triangle on triangle offset type may be the logical choice. An example of such an application is the hangar at Frankfurt-on-Main Airport, which has a span of 83.198 m. The distances between two adjacent nodes in two orthogonal directions defines the grid module. To achieving a cost-effective solution, the grid module has to be chosen to be as large as practicable in order to minimize the number of node connectors, which are an expensive item. Ideally, purlins receiving loads from the roof cladding must transfer the loads to the space frame only at the nodes. If this is the case, the members of the space frame will carry only axial loads. This implies the placing of purlins only at the nodes and that the roof sheeting must be strong enough to span between the purlins so placed. In other words, the roof cladding chosen determines the spacing of purlins, which in turn determines the length of the grid module. In Europe and the Middle East, steel and aluminium troughed sheets designed to span 3 m are available off the shelf and it is logical to adopt a 3 m × 3 m module. In India the maximum possible span with readily available troughed aluminium or colour-coated steel sheets is 2 m. It is, of course, possible to specially order sheets that can span 3 m, but they tend to be expensive. Hence there is a trade-off between using 3 m grids with more expensive sheets, and using intermediate purlins resting directly on members and the consequent need to design them for both bending and axial loads.

## 4.2. Case study: sports hall roof, Al Wahda Sports Club, Abu Dhabi

We present here the design of a space frame for the sports hall of the Al Wahda Sports Club, Abu Dhabi. The problems that arose and how they were addressed are explained.

### *4.2.1. Notes on the project*

The plan, longitudinal section and cross-section of the Al Wahda Sports Hall designed by the architects Weightman, Bullen and Taylor International for the Al Wahda Municipality are shown in Fig. 4.1. Designed for handball, tennis and basketball, the floor plan provides for a playing field of 54 m × 36.2 m and a grand-stand with stepped seats occupying 54 m × 17.1 m. The overall rectangular plan area to be covered was 54 m × 43.30 m.

### *4.2.2. Architectural constraints*

The space frame was to be supported on reinforced concrete columns spaced 6 m apart along the longer sides and 300 mm thick reinforced concrete walls along the shorter sides. The support points had already been selected by the structural consultants R. J. Crocker and Partners. In fact, the columns and the walls were already in place before bids were invited from specialist contractors for the design, fabrication and supply of the components of the steel space frame. The clear height had been specified as 8.35 m. With this constraint, the structural depth of the space frame became restricted to 2.25 m. A higher structural depth would have been advantageous but was not available. In these circumstances, it was necessary to work from bottom to top instead of from top to bottom. We had to make do with the columns and walls available and restrict the vertical and horizontal reactions transferred to them by the space frame to what they could safely carry. In short, we were being called upon to design a head to suit an available cap! A situation of this kind is totally avoidable if the architect, the structural consultant and the specialist on space frames are on the scene from the very beginning. Such interaction during the conceptual phase of a project will lead to a rational choice of modules, structural

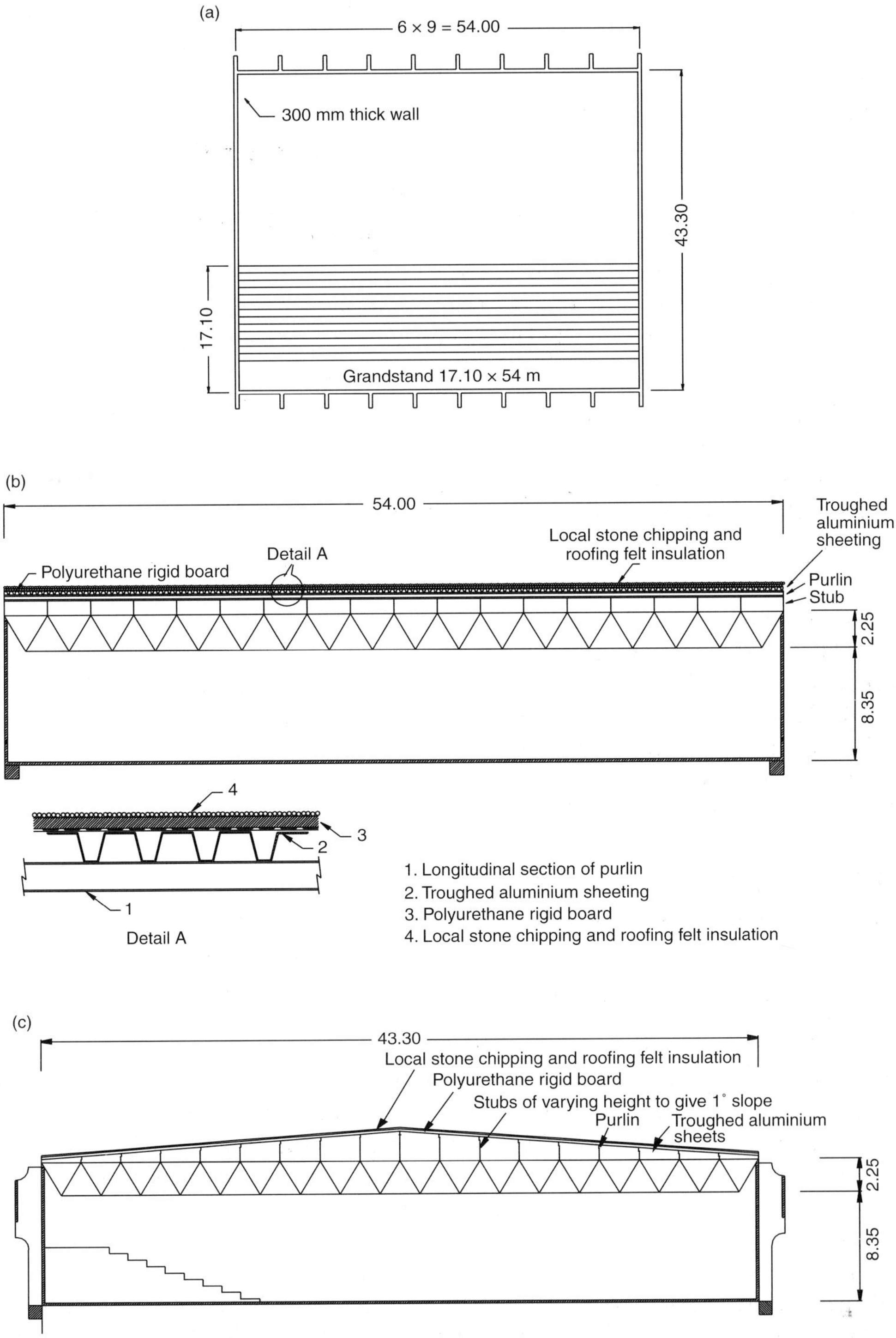


*Fig. 4.1. The indoor stadium of Al Wahda Sports Club: (a) ground plan; (b) longitudinal section; (c) cross-section*

depth and topology, as well as to proper integration of services in the space frame, as discussed in a later section.

### 4.2.3. *Creation of a consortium*

At the initiative of the first author, a multi-national space-frame building consortium was set up to execute the project. The division of labour among the partners was as follows:

- Structural analysis and design was carried out by the Civil Engineering Consultancy Consortium of Chennai, India, headed by the first author, and vetted by Octatube Space Structures BV, The Netherlands, headed by Dr Mick Eekhout.
- Flocoat tubes of 450 MPa yield strength were fabricated by Nagarjuna Coated Tubes at their factory at Hyderabad, India, and shipped to Dubai.
- The Tuball spherical node connectors were supplied by Octatube Space Structures BV, The Netherlands.
- Assembly and erection of the space frame was by Galadari Engineering Works of Dubai under the direction of Consolidated Construction Company, who were the general contractors for the job.

### 4.2.4. *Strategy for winning the contract*

A multi-pronged strategy was worked out to win the contract in the face of stiff international competition. The factors that eventually tilted the award of the contract in our favour were:

- The design using Flocoat tubes of 450 MPa yield strength had an edge over the European designs using tubes of 240 MPa yield strength. This factor helped in pruning the weight of the space frame.
- A further reduction in weight was possible by adopting a square over diagonal topology, which resulted in a drastic reduction in the number of diagonal bracing members and a consequent reduction in weight.
- The space frame proposed by us weighed only 50 t, compared with 80 t of the space frame proposed by our nearest competitor.
- Flocoat tubes have a superior corrosion resistance due to the rust-resistant treatment that they undergo on-line during manufacture.
- There was a substantial reduction in the number of node connectors required, due to the adoption of a square over diagonal topology that required bottom nodes only in alternate bays

### 4.2.5. *Selection of topology*

The square on square offset, set diagonally, topology was the first choice of the authors, because it would involve only 15 kg/m$^2$ of steel. This configuration is shown in Fig. 4.2. However, this topology was ruled out by the general contractor because it would involve the provision of an extra reinforced concrete beam connecting the

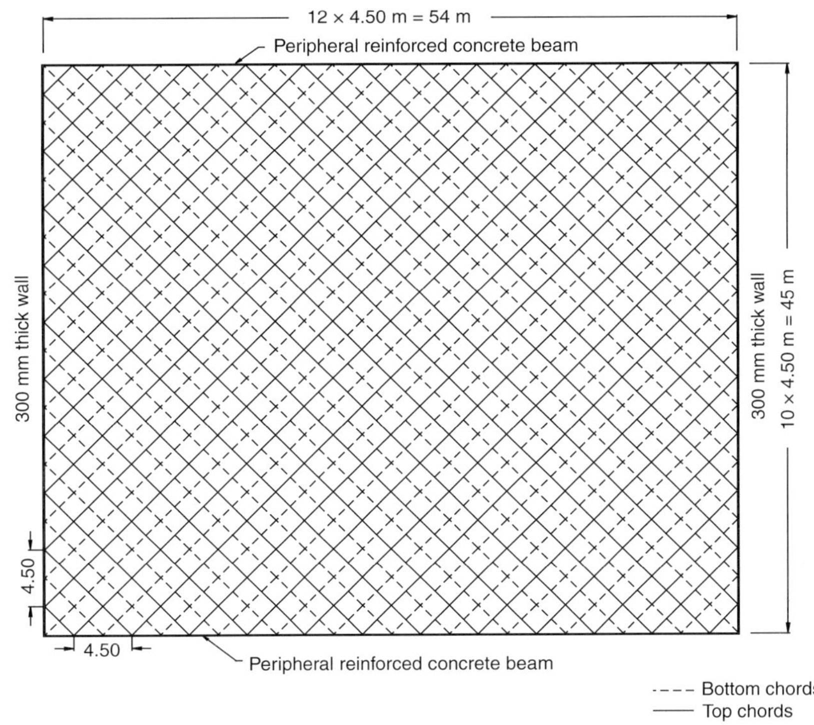


*Fig. 4.2. Square on square offset, set diagonally topology*

existing columns to permit the nodes to be seated at 4.5 m intervals along the longer side. The available columns were 6 m apart. We were therefore obliged to adopt the square on diagonal configuration.

### 4.2.6. Tubes

Flocoat tubes with a yield strength of 450 MPa, manufactured by Nagarjuna Coated Tubes, Hyderabad, India, were used for the space frame members. They were powder-coated at Dubai.

### 4.2.7. Node connectors

Tuball spherical node connectors of 162 mm diameter with a breaking strength of 800 kN were used for this project. The solid props at the end of the tubes had a yield strength of 450 MPa to match the strength of the tubes. High-strength, hot-dip galvanized bolts of grades 8.8, 10.9 and 12.9 conforming to the relevant British Standard specifications were used in the connectors. The bolts had either hexagonal or Allen-type heads. Where members carried high tensile forces in excess of what the nodes could carry, threaded bolts of the same quality were run through the connectors. The cap of the Tuball node is normally secured to the cup by means of an M12 grade Allen-head bolt. When threaded rods were run through the connector, the Allen-head bolt was replaced by two Parker screws or an umbrella bolt was hooked onto the rod.

### 4.2.8. Loads

The space frame was designed for a vertical load of 1.5 kN/m$^2$, the break-up of which is as follows:

Dead load including self-weight 0.50 kN/m$^2$
Live load 0.50 kN/m$^2$
Weight of hung fixtures and air-conditioning ducts 0.50 kN/m$^2$
Total 1.50 kN/m$^2$

### 4.2.9. Codes

The contract required BS 5950: Part 1: 1985 to be generally followed. Eurocode 3: 1989 provisions were also acceptable wherever they were found to be appropriate.

### 4.2.10. Effective load factor

BS 5950: Part 1: 1985 specifies a load factor of $\gamma_f = 1.40$ for dead loads and 1.60 for live loads. Hence the design was for an effective load factor of

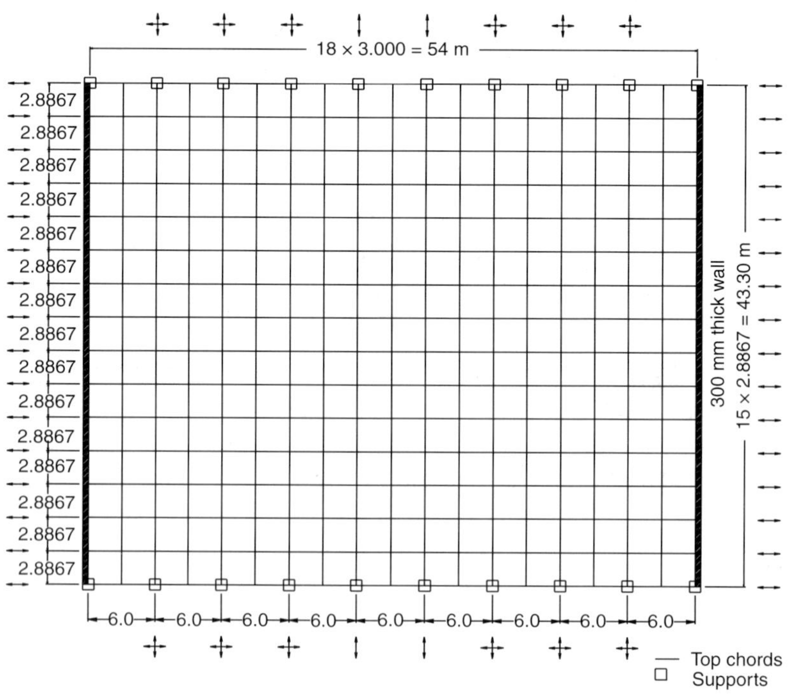


*Fig. 4.3. General plan of the space frame, showing the positions, dimensions and restraints of the columns*

$$\gamma_{\mathrm{f}} = \frac{1.4 \times 1 + 1.6 \times 0.5}{1.50} = 1.47$$

### 4.2.11. Layout

The general layout plan of the space frame, showing the column locations, dimensions and boundary conditions, is given in Fig. 4.3. The plans of the top chords, bottom chords and bracing diagonals are shown in Figs 4.4 to 4.6, which also give the node and member numbering.

### 4.2.12. Boundary conditions

The space frame was supported on the top nodes. To minimize thermal stresses, movements of the nodes over the supports were permitted as shown in Fig. 4.3.

### 4.2.13. Equivalent lengths

The equivalent lengths of the compression members were assumed to be 0.85 of their actual lengths.

### 4.2.14. Method of analysis

The loads were multiplied by the effective load factor to arrive at the value of the loads at the ultimate limit state. These loads were applied at the top nodes and a linear computer analysis was carried out using SAP90 software to determine the forces in the members, the displacements at the nodes and the reactions at the supports.

### 4.2.15. Design of the tension and compression members

To facilitate the design of the tension and compression members, the permissible tension and compression in the tubes in accordance with BS 5950: Part 1: 1985 were computed. These values are given in Table 4.1.

### 4.2.16. Selection of preliminary sizes for the first cycle of the analysis

Regarding a strip of the space frame between adjacent nodes as a beam, a free body diagram can be drawn as shown in Fig. 4.7. The distribution of the total load (2.25 kN/m$^2$) in the $x$ and $y$ directions is inversely proportional to the fourth powers of $l_x$ and $l_y$, such that

$$p_x = \frac{pl_y^4}{l_x^4 + l_y^4} \tag{4.1a}$$

and

$$p_y = \frac{pl_x^4}{l_x^4 + l_y^4} \tag{4.1b}$$

*Fig. 4.4. The node and member numbering for the top chords. (Node numbers (next to the circles) 1–304; member numbers (on the connecting lines) 1–573)*

*Fig. 4.5. The node and member numbering for the bottom chords. (Node numbers (in the circles) 305–470; member numbers (on the connecting lines) 574–873)*

*Fig. 4.6. The node (circles) and member numbering for the web members. (Node numbers (on right-hand side) 1–470; member numbers (on the connecting lines) 874–1537)*

*Table 4.1. Allowable tension and compression in the tubes at the ultimate limit state*

| Tube size (mm) | Area (mm$^{2)}$ | $r$ (mm) | $l_e/r$ | Permissible compression (kN) | Permissible tension (kN) |
|---|---|---|---|---|---|
| *Length 3000 mm; effective length,* $l_e = 0.85 \times 3000 = 2550$ *mm* | | | | | |
| 139.70 × 3.65 | 1561 | 48.10 | 53.00 | 600 | 702 |
| 127.00 × 3.65 | 1415 | 43.60 | 58.50 | 515 | 637 |
| 114.30 × 3.65 | 1269 | 39.10 | 65.20 | 424 | 571 |
| 101.60 × 3.65 | 1124 | 34.60 | 73.60 | 327 | 506 |
| 76.20 × 3.65 | 832 | 25.70 | 99.20 | 151 | 375 |
| *Length* $= 2886.7$ *mm; effective length,* $l_e = 0.85 \times 2886.7 = 2453.7$ *mm* | | | | | |
| 139.70 × 3.65 | 1561 | 48.10 | 51.10 | 610 | 702 |
| 127.00 × 3.65 | 1415 | 43.60 | 56.28 | 528 | 637 |
| 114.30 × 3.65 | 1269 | 39.10 | 62.75 | 439 | 571 |
| 101.60 × 3.65 | 1124 | 34.60 | 70.92 | 343 | 506 |
| 76.20 × 3.65 | 832 | 25.70 | 95.47 | 162 | 375 |
| *Length* $= 4163.30$ *mm; effective length,* $l_e = 0.85 \times 4163.30 = 3539$ *mm* | | | | | |
| 139.70 × 3.65 | 1561 | 48.10 | 73.58 | 454 | 702 |
| 127.00 × 3.65 | 1415 | 43.60 | 81.17 | 359 | 637 |
| 114.30 × 3.65 | 1269 | 39.10 | 90.51 | 270 | 571 |
| 101.60 × 3.65 | 1124 | 34.60 | 102.28 | 194 | 506 |
| 76.20 × 3.65 | 832 | 25.70 | 137.70 | 83 | 375 |
| *Length* $= 3065.20$ *mm; effective length,* $l_e = 0.85 \times 3000 = 2605.42$ *mm* | | | | | |
| 139.70 × 3.65 | 1561 | 48.10 | 54.17 | 594 | 702 |
| 127.00 × 3.65 | 1415 | 43.60 | 59.76 | 508 | 637 |
| 114.30 × 3.65 | 1269 | 39.10 | 66.63 | 415 | 571 |
| 101.60 × 3.65 | 1124 | 34.60 | 75.30 | 318 | 506 |
| 76.20 × 3.65 | 832 | 25.70 | 101.38 | 146 | 375 |

where $p$ is the total load and $p_x$ and $p_y$ are the shares of the loads carried by beam strips in the $x$ and $y$ directions.

Hence,

$$p_x = \frac{pl_y^4}{l_x^4 + l_y^4} = 2.25\left(\frac{54^4}{43.3^4 + 54^4}\right) = 1.59 \text{ kN/m}^2$$

The top grid will function as a beam on an elastic foundation, the elastic foundation being provided by the bottom grid.

Thus, assuming partial fixity,

$$T = \frac{1.59 \times 43.30^2}{10}\,\frac{3}{2.25} = 397 \text{ kN}$$

$$\text{C} = \frac{T}{\sqrt{2}} = \frac{397}{\sqrt{2}} = 281 \text{ kN}$$

We therefore adopted a preliminary tube size of 114.30 mm × 3.65 mm for the top chords. It can be seen from Table 4.1 that this tube can carry a permissible

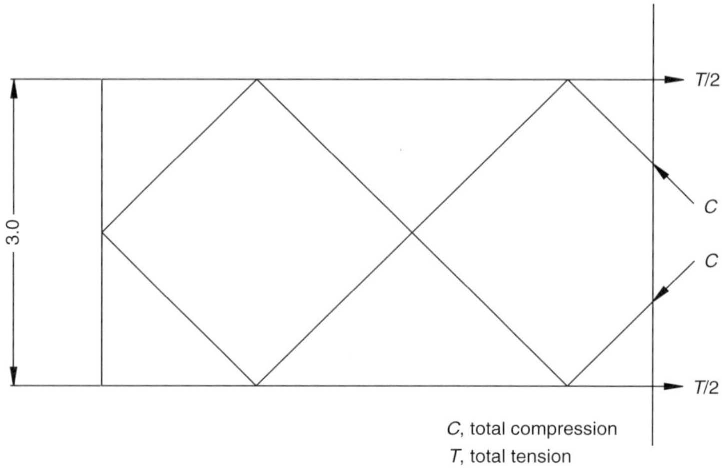


*Fig. 4.7. Free body diagram of a strip of the space frame between adjacent nodes*

compression of 424 kN, which is greater than the required value of 281 kN. A tube of size 101.60 mm × 3.65 mm with a permissible tension of 506 kN was adequate for the bottom chords to carry a tension of 397 kN. There is no simple rule for selecting the preliminary size of the diagonal bracing. Based on past experience a size of 76.10 mm × 3.65 mm was assumed. The SAP90 analysis was then carried out using these preliminary sizes.

### *4.2.17. Computer output*

The computer output consisted of the forces in the members, the displacements at the nodes and the reactions at the supports. The input data and computer output are available on the accompanying CD-ROM. Wherever the forces in the members exceed the capacity of the member sizes initially chosen, the member size was increased. The member sizes chosen to match the forces are given in Figs 4.8 to 4.10.

The maximum compression of 615 kN occurs in top chord member No. 136. The maximum tension of 511 kN occurs in bottom chord member No. 662. The maximum compression of 151 kN occurs in diagonal bracing member No. 1204, and the maximum tension of 280 kN occurs in diagonal bracing member No. 394.

### *4.2.18. Check for deflection*

The maximum deflection of 494 mm occurs at node 388 under loads corresponding to the ultimate limit state. At the serviceability limit state, the deflection was computed to be 494/1.47 = 336 mm. The unfactored imposed loads consist of the live load of 0.5 kN/m$^2$ and the weight of the air-conditioning ducts, estimated at 0.033 kN/m$^2$, which together add up to 0.533 kN/m$^2$. Hence the deflection under imposed loads at the serviceability limit state is (0.533/1.50)336 = 119 mm. This is less than $l$/200 (i.e. 43 300/200 = 216.50 mm), as permitted by BS 5950: Part 1: 1985 under the action of imposed loads.

*Fig. 4.8. The node numbers and member sizes for the top chords. (Node numbers (next to the circles) 1–304. Members: (1) 114.30 × 3.65 mm; (2) 127.00 × 3.65 mm; (3) 139.70 × 3.65 mm; (4) 152.40 × 3.65 mm)*

*Fig. 4.9. The node numbers and member sizes for the bottom chords. (Node numbers (in the circles) 305–470. Members: (1) 101.60 × 3.65 mm; (2) 114.30 × 3.65 mm)*

*Fig. 4.10. The node numbers and member sizes for the bracing members. (Node numbers (in the circles) 1–304. Members: (1) 76.10 × 3.65 mm; (2) 101.60 × 3.65 mm)*

### *4.2.19. Support design*

From the support reactions shown in Table 4.2 it is possible to recognize three types of support, designated as $A_1$, $A_2$ and B. At type $A_1$ supports there are reactions in the $x$ and $z$ directions. At $A_2$ joints there are only vertical reactions. At type B supports reactions act in the $y$ and $z$ directions.

By way of illustration, we describe here the design of the type $A_1$ support at node 11, where there is a horizontal reaction of 352 kN in the $x$ direction and a vertical reaction of 260 kN. The node is supported on a 76.10 mm × 6.30 mm tube stiffened with 25 mm thick triangular ribs and welded to a 200 m × 350 mm anchor plate with four holes for grade 8.8 anchor bolts (Fig. 4.11).

*Check for the compression in the tube*
Vertical force = 260 kN
Area of tube = $(76.10 - 6.30)(\pi \times 6.30) = 1381\ \text{mm}^2$
Compressive stress = $(260\,000/1381) = 188$ MPa

The calculated value is less than 450 MPa, and hence a tube of this size is safe.

*The tension in the anchor bolts*
Bending moment = $(352\,000)(300) = 105.60 \times 10^6$ N-mm
Hence the tension in the anchor bolt due to bending is

$$\frac{105.60 \times 10^6}{2 \times 250} = 211\ 200\ \text{N} = 211\ \text{kN}$$

Compression in bolt = 260/4 = 65 kN
Hence the tension in the bolt is 211 – 65 = 146 kN
The proof strength of a grade 8.8 bolt is 640 MPa, and thus the area required is $146\,000/640 = 228\ \text{mm}^2$

M20 bolts with an area of 245 $\text{mm}^2$ are therefore sufficient.

*Check of the junction between the tube and the anchor plate*
We need to calculate the moment of inertia about the $yy$ axis:

Contribution of the tube = $840\,489\ \text{mm}^4$
Contribution of the vertical ribs
$= (1/12)(25 \times 2)(175 - 38)^3 + [(350 - 76)/2](25 \times 2 \times 106.5)^2$
$= 10\,713\,971 + 77\,694\,413 = 88\,408\,384\ \text{mm}^4$
Contribution of the horizontal ribs = $2(100 - 38)(1/12)25^3 = 161\,458\ \text{mm}^4$
Total moment of inertia about the $yy$ axis = $89\,410\,332\ \text{mm}^4$
Modulus of the section about the $yy$ axis = $89\,410\,332/175 = 510\,916$ mm

*Bending stress at the junction*
Between the tube and the anchor plate = $(105 \times 10^6)/510\,916 = 206$ MPa
The yield strength of the material of the anchor plate and ribs is 240 MPa, and hence the junction is safe.

*Table 4.2. Support reactions*

| Joint | $F(x)$ | $F(y)$ | $F(z)$ | Support type |
|---|---|---|---|---|
| 1 | 0 | 56.7178 | –88.4170 | B |
| 3 | 0 | 0 | 91.2397 | $A_2$ |
| 5 | 0 | 0 | 172.2487 | $A_2$ |
| 7 | 0 | 0 | 212.4564 | $A_2$ |
| 9 | 351.9947 | 0 | 256.2400 | $A_1$ |
| 11 | –351.9947 | 0 | 260.2154 | $A_1$ |
| 13 | 0 | 0 | 212.3254 | $A_2$ |
| 15 | 0 | 0 | 174.7464 | $A_2$ |
| 17 | 0 | 0 | 80.2009 | $A_2$ |
| 19 | 0 | 52.0745 | –81.1788 | B |
| 20 | 0 | 64.6473 | 16.1685 | B |
| 38 | 0 | 30.3855 | 65.2188 | B |
| 39 | 0 | 20.0956 | 53.2827 | B |
| 57 | 0 | 75.1473 | –27.6647 | B |
| 58 | 0 | 106.9806 | 42.0098 | B |
| 76 | 0 | 27.9958 | 101.1687 | B |
| 77 | 0 | 12.0431 | 105.9875 | B |
| 95 | 0 | 120.4265 | 60.6954 | B |
| 96 | 0 | 119.6283 | 101.5229 | B |
| 114 | 0 | –1.8202 | 129.8740 | B |
| 115 | 0 | –29.5975 | 131.1039 | B |
| 133 | 0 | 105.6490 | 114.8789 | B |
| 134 | 0 | 83.7351 | 130.5340 | B |
| 152 | 0 | –55.2852 | 136.0000 | B |
| 153 | 0 | –83.7351 | 130.5340 | B |
| 171 | 0 | 55.2852 | 136.0000 | B |
| 172 | 0 | 29.5975 | 131.1039 | B |
| 190 | 0 | –105.6490 | 114.8789 | B |
| 191 | 0 | –119.6283 | 101.5229 | B |
| 209 | 0 | 1.8202 | 129.8740 | B |
| 210 | 0 | –12.0431 | 105.9875 | B |
| 228 | 0 | –120.4265 | 60.6954 | B |
| 229 | 0 | –106.9806 | 42.0098 | B |
| 247 | 0 | –27.9958 | 101.1687 | B |
| 248 | 0 | –20.0956 | 53.2827 | B |
| 266 | 0 | –75.1473 | –27.6647 | B |
| 267 | 0 | –64.6473 | 16.1685 | B |
| 285 | 0 | –30.3855 | 65.2188 | B |
| 286 | 0 | –56.7178 | –88.4170 | B |
| 288 | 0 | 0 | 91.2397 | $A_2$ |
| 290 | 0 | 0 | 172.2487 | $A_2$ |
| 292 | 0 | 0 | 212.4564 | $A_2$ |
| 294 | 351.9947 | 0 | 256.2400 | $A_1$ |
| 296 | –351.9947 | 0 | 260.2154 | $A_1$ |
| 298 | 0 | 0 | 212.3254 | $A_2$ |
| 300 | 0 | 0 | 174.7464 | $A_2$ |
| 302 | 0 | 0 | 80.2009 | $A_2$ |
| 304 | 0 | –52.0745 | –81.1788 | B |

*Check for shear*
Area of M20 bolt = 245 mm$^2$
Allowable shear stress of grade 8.8 bolt = 498 MPa
Shear strength of four M20 bolts = $(4 \times 498 \times 245)/1000 = 488$ kN

The calculated value is greater than the required value of 353 kN, and hence the bolts are safe.

*Check for bearing*
Assume the anchor plate to be 25 mm thick
Tensile strength of anchor plate, $U_s = 360$ MPa
Yield strength of anchor plate, $Y_s = 240$ MPa
The permissible strength in the bearing according to Table 33 of BS 5950: Part 1: 1985 is $0.65(U_s + Y_s) = 0.65(360 + 240) = 390$ MPa

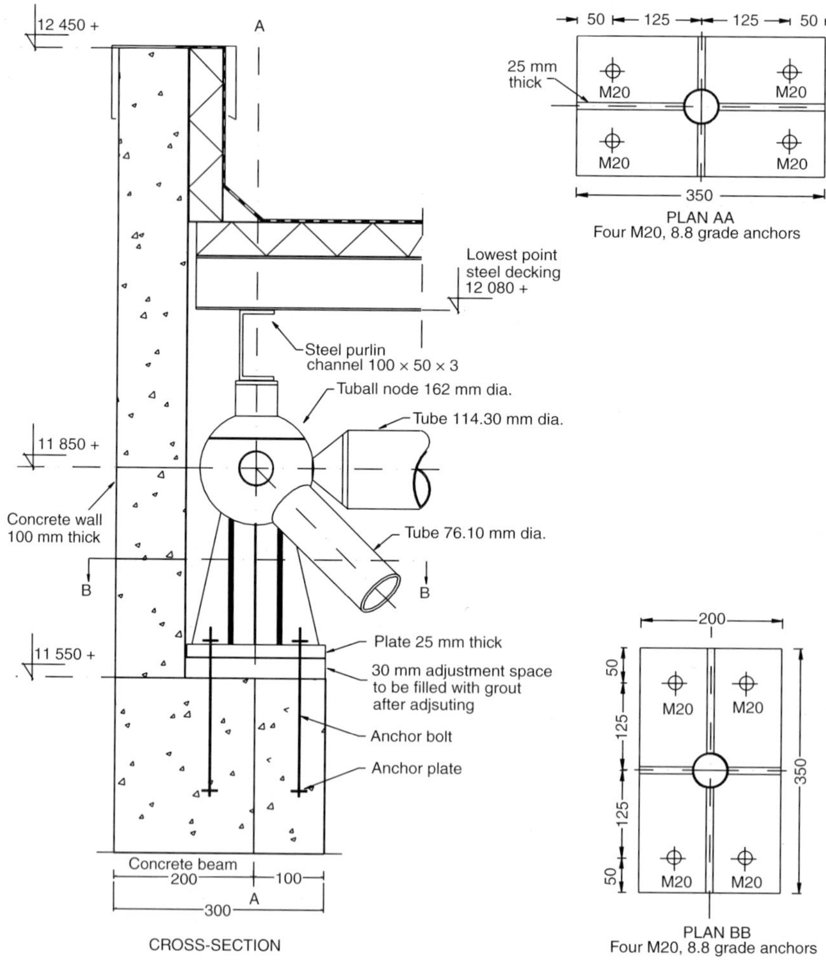


*Fig. 4.11. Details of the support at node 11*

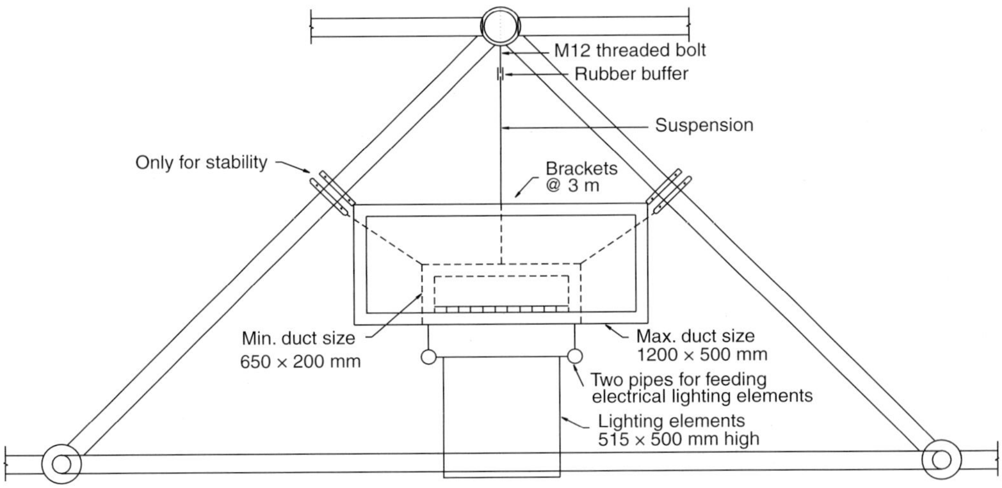


*Fig. 4.12. Details of the air-conditioning and lighting*

Hence, the resistance against bearing is $P_{bs} = (4 \times 20 \times 25 \times 390)/1000 = 780$ kN

This is greater than the required value of 352 kN, and hence the anchor plate is safe.

*Design of the Teflon pad*
Material properties:

Thickness 1.5 mm
Tensile/compressive strength 18 MPa
Elongation permissible 300%
Young's modulus of elasticity 400 MPa
Bending strength 18.30 MPa

The working stress method will be used for the design of the Teflon pad:

Vertical reaction at the serviceability limit state = 260/1.47 = 177 kN
Compression in the Teflon pad = $(177 \times 1000)/(350 \times 200) = 2.52$ MPa (i.e. less than 18 MPa)
Axial compression of the pad = $(2.52 \times 1.5)/400 = 0.00945$ mm
Contraction = $(0.00945 \times 100)/1.5 = 0.63\%$

The design of the Teflon pad is therefore suitable.

## 4.3. Services

The service contractor had initially planned to install the air-conditioning ducts and lighting fixtures after the space frame had been hoisted and placed in position. At the suggestion of the second author, it was decided to install these fixtures in the space frame while it was still on the ground in order to avoid the hazards and inconvenience of working at heights. The arrangement worked out for housing the services within the space frame is shown schematically in Fig. 4.12.

## 4.4. Assembly and hoisting

The dimensions of the space frame are larger than the inner dimensions of the reinforced concrete supporting structures, because the frame was designed to be supported on its top nodes. The presence of the sloping grandstand ruled out the assembly of the complete space frame on the floor of the stadium. Normally, space frames are assembled at ground level on the concrete floor and hoisted up almost vertically by mobile cranes to be placed on columns and walls. For the reasons already mentioned this procedure was not feasible and an alternative scheme had to be worked out. The sequence adopted is shown schematically in Fig. 4.13. Module A, measuring 43.30 m × 30 m, was assembled on the ground floor of the stadium. Its position inside the stadium is shown by the dotted lines in Fig. 4.13. It was hoisted, taken outside the stadium, turned round by 270°, brought inside the stadium and placed in its final position (indicated by the solid lines in Fig. 4.13). Module B was assembled outside the stadium, hoisted and placed in its final position. A 350-t crane with a 70-m jib was assembled outside the building to carry out the hoisting operation. Special precautions had to be taken to reinforce the diagonal lower chord member squares at their open ends in order to prevent them from deforming into rhombuses. These squares were provided with temporary tensile elements, 6 m long, inserted in the direction in which the squares were predicted to elongate. Because of the open end of the space frame, which did not have a closed geometry, the upward deflection during hoisting over a distance of 43.30 m was computed to be in the region of 350–400 mm. This in fact proved to be the case. The three closed sides,

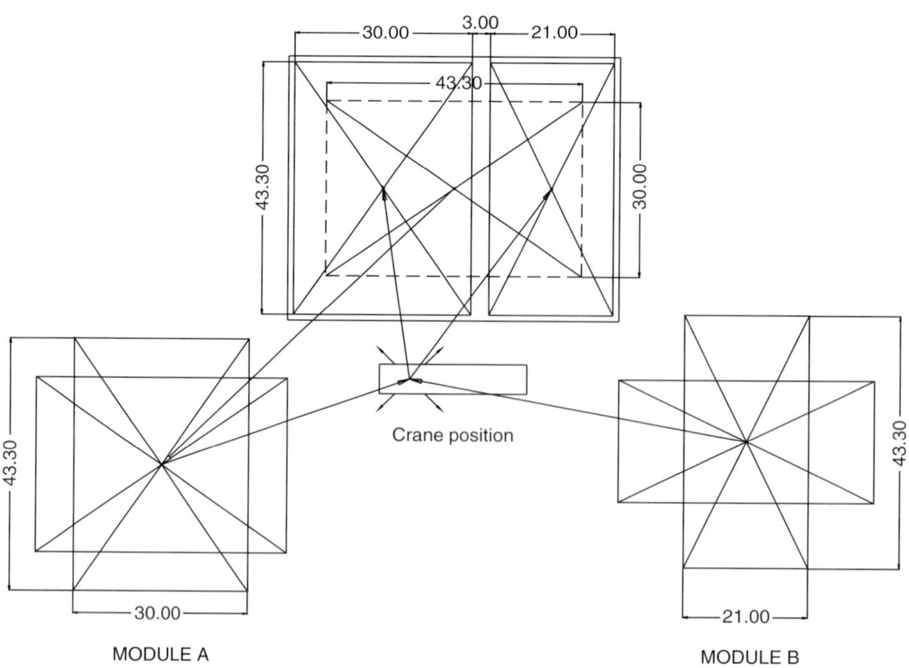


*Fig. 4.13. The erection sequence for the space frame*

*Fig. 4.14. The space frame being hoisted*

being more rigid, did not need any reinforcement. Several hoisting positions at several cable angles were computed. The open ends of modules A and B, which were 3 m apart, were supported on scaffolding towers. The scaffolding props were withdrawn after the space frame had been completed by installing the members in the gap. Figure 4.14 shows the space frame being hoisted by a 300-t crane.

# 5. Time- and labour-saving aids for pre- and post-processing tasks

## 5.1. Input data generation

The sea change brought about by computers in the realms of structural analysis, design and optimization had not been matched until recently by parallel developments in input data generation, which continues to be a tedious, time-consuming and error-prone chore. Nooshin [5.1] succinctly summarized this position when he observed that there is only one task that is more laborious and boring than the generation of data, that is checking the data! These considerations led him to invent Formex algebra in its embryonic form in 1975. These ideas were further developed and published in 1984 [5.2]. This was soon followed up by the development of Formian, an interactive programming language for use in conjunction with Formex concepts, which allows data manipulation of configurations on runtime. These breakthroughs have ushered in a revolution in automatic data generation.

## 5.2. Formex data generation

The advantages offered by the Formex formulation as a tool for data generation have been summed up by Nooshin as follows:

- The process of data generation for structural systems is vastly simplified. The more complex a structure is, the more advantageous it is to resort to Formex formulation.
- The reduction in storage requirements can be enormous. For large structures, the storage required for data in explicit form is several hundred times that needed for data stored in the Formex format. The Formex formulation of input data for a space frame hangar with 3514 nodes and 12 926 members can be neatly compressed into a few pages of ASCII text [5.3]. There can be no better demonstration of the power and elegance of the Formex formulation.
- The formulation may be modified as the design proceeds as easily as modifying a document on a word processor, and it can be manipulated to reflect the current

status after making any necessary changes in geometry, support conditions or loading.

- It is not necessary to draw a configuration in order to generate the data for that configuration. However, once the Formex formulation is available, the configuration can be viewed on a monitor or printed.
- An actual or implied node numbering system is not necessary either for a Formex description of a configuration or for carrying out stress analysis by using the input data in a compatible stress analysis package. By using the Lineation function [5.4] it is be possible to obtain a graphical output of the node numbers, member numbers, element types, loads and forces. Alternatively, a drafting package such as Autocad or Microstation may be used.
- Formex data generation has a distinct edge over other available schemes that are analysis package dependent and hence lack versatility. On the other hand, the input data generated by the Formex formulation can be used, via Formian, in a variety of analysis packages, such as ABAQUS, SAP2000 and LUSAS.

## 5.3. Pre- and post-processing tasks in the analysis and design of space frames

The analysis and design phases of a space frame involve the following sequences:

*Pre-processing*

- choosing an appropriate module
- selecting the geometrical configurations for different layers
- defining the support conditions
- prescribing the loads and load combinations and determining the loads to be applied at the nodes
- selecting the member sizes for the preliminary run
- generating the topology of each layer and determining the co-ordinates of the nodes
- numbering the nodes and members.

*Stress analysis and optimization*

- selecting the member sizes to the match forces
- optimizing the design to the minimum weight.

*Post-processing*

- displaying the forces and deformed geometry graphically.

The Formex formulation and Formian are fully geared to meet all these requirements.

## 5.4. Formex and Formian fundamentals

Two systems of dotted orthogonal lines, known as *normats*, define the reference system (Fig. 5.1). The intersections of these lines are called *normat points*. The construct [1, 0; 0, 1] representing an element *ij* is a simple example of a Formex. The

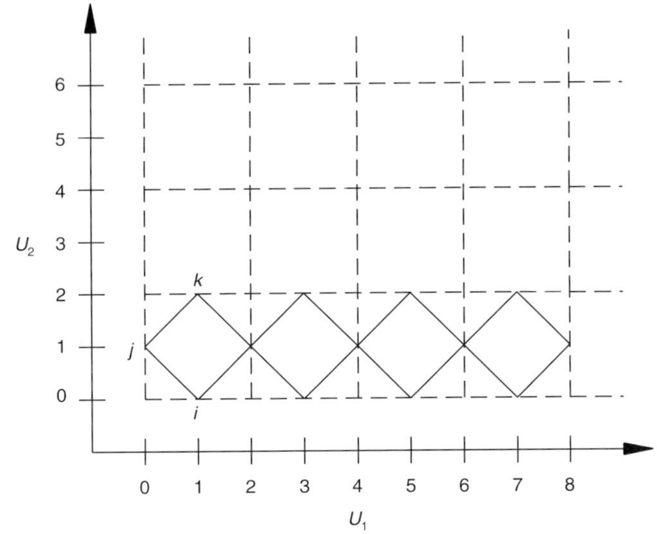


*Fig. 5.1. The Formex formulation*

equation $E_1 = [1, 0; 0, 1]$ would mean that the value of the variable $E_1$ is given by the Formex on the right-hand side. Here 1, 0, 0, 1 are *uniples* and 1, 0 and 0, 1 are *signets*. Similarly, the element *jk* can be represented by the equation $E_2 = [0, 1; 1, 2]$ (see Fig. 5.1). The two elements *ij* and *jk* may together be represented by the Formex variable $F = \{[1, 0; 0, 1], [0, 1; 1, 2]\}$. Here [1, 0; 0, 1] and [0, 1; 1, 2] are called *cantles*.

Shorn of technical jargon, the Formex formulation is an elegant tool for pattern composition. Repetitive patterns are easily generated by Formian functions such as the *Rindle function* RIN and the *lamda function* LAM. The Rindle function, which gives 'translational replication', takes the form

$$H = \text{RIN}(1, 4, 2)\,|\,F; \tag{5.1}$$

where 1 is the direction of replication, 4 is the number of replications and 2 is the amount of translation at each step. The rallus symbol stands for 'of'.

The lamda function gives 'reflectional translation' and is stated as

$$R = \text{LAM}(1, 2)\,|\,F; \tag{5.2}$$

where 1 is the direction of replication and 2 is the position of the plane of reflection.

Referring to Fig. 5.1 the pattern represented by the variable $K$ is the result of the reflection of $F$ at plane 1 and replication by translation four times. Thus,

$$K = \text{RIN}(1, 4, 2)\,|\,\text{LAM}(1, 1)\,|\,F; \tag{5.3}$$

When replication by translation in two directions is desired, the more powerful *Tendid Rindle function* RIND may be used. For example, in RIND(10, 6, 2, 3), 10 is the number of replications in the first direction, 6 is the number of replications in the second direction, 2 is the amount of translation in the first direction and 3 is the degree of translation in the second direction.

Similarly, the *lamda effect in two directions* function LAMID may be employed with advantage when replication by reflection of a pattern in two directions is involved. Thus, in the statement LAMID(1, 1) the first number in the brackets denotes the plane of reflection in the first direction and the second number denotes that in the second direction.

Application of these Formian functions is illustrated below, where we generate the top and bottom chords for the flat double-layer space frame shown in Fig 5.2. The Formex formulation takes the following form:

$$\text{TOP} = \text{RINID}(4, 2, 2, 2)\,|\,\text{LAMID}(1, 2)\,|\,\{[1, 0, 1; 0, 1, 1], [0, 1, 1; 1, 2, 1]\}; \tag{5.4}$$

$$\text{BOT} = \text{RINID}(4, 3, 2, 2)\,|\,\text{LAMID}(1, 1)\,|\,\{[0, 0, 0; 1, 0, 0], [0, 0, 0; 0, 1, 0]\}; \tag{5.5}$$

For a more detailed account of Formian functions, reference may be made to the Formian manual [5.4].

## 5.5. Compretic and normic properties

The manner in which the configuration is divided into component parts and the way in which the component parts are interrelated constitute the *compretic properties*. Nodal co-ordinates defining the position of points in a configuration are referred to as *normic properties*. The collection of the compretic properties of a configuration is known as the *compret* of the configuration.

A rule for the transformation of signets of Formex into geometric points is referred to as a *retronorm*. The uniples $[U_1, U_2, U_3]$ of a signet can be converted into Cartesian co-ordinates $x, y, z$ by the use of the *basitrifect retronorm*, which takes the form of the following three equations:

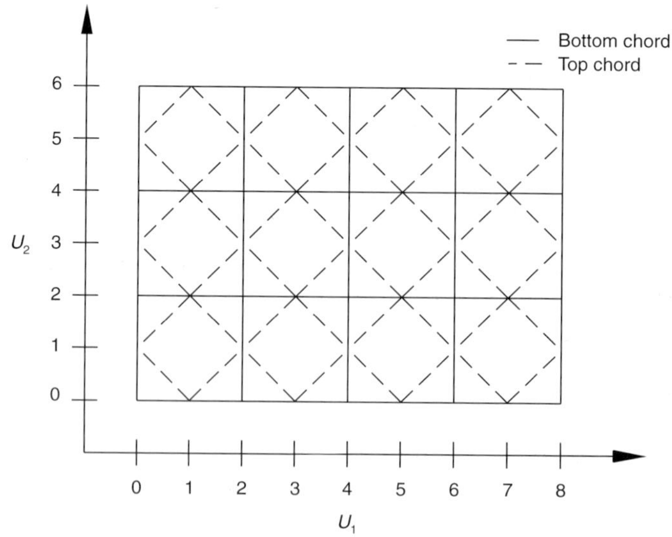


*Fig. 5.2. The top and bottom chords for the example flat double-layer space frame*

$$x = b_1 * U_1 \tag{5.6}$$

$$y = b_2 * U_2 \tag{5.7}$$

$$z = b_3 * U_3 \tag{5.8}$$

where $b_1$, $b_2$ and $b_3$ may be regarded as scaling functions.

## 5.6. Cylindrical transformations

Cylindrical transformations are illustrated by means of their application to two examples of barrel vaults.

*Example 5.1*

The example chosen is that of a braced barrel vault, stabilized by triangulated diaphragms at the ends, to cover a tennis court for the SNCF at Paris–Vaugirard, designed by Du Château [5.5]. The Formex formulation closely follows that outlined by Yassaee [5.6]. The barrel vault is semi-circular, with a chord width of 18 m, length of 36 m and height 9 m (Fig. 5.3). The axis $U_1$ is chosen to lie along the radius and its positive direction is radially outward from the centre. The $U_2$ axis lies along the circumference and $U_3$ lies along the length of the barrel. The configuration is first generated in terms of the uniples $U_1$, $U_2$ and $U_3$ and this is later transformed into cylindrical co-ordinates.

*Bracing diagonals*

$$C_1 = \text{RINIT}(9, 20, 4, 2)|\text{LAMIT}(2, 1)|[9, 0, 0; 9, 2, 1]; \tag{5.9}$$

The arguments of RINIT indicate the number of repetitions in the directions $U_2$ and $U_3$ and the amount of translation at each step. The arguments of LAMIT denote the positions of the planes at which reflection takes place.

*Elements parallel to the generatrix of the cylinder*

$$C_2 = \text{RINIT}(10, 20, 4, 2)|[9, 0, 0; 9, 0, 2]; \tag{5.10}$$

$$C_3 = \text{RINIT}(9, 19, 4, 2)|[9, 2, 1; 9, 2, 3]; \tag{5.11}$$

$$C_4 = \text{LAM}(3, 20)|\text{RIN}(2, 9, 4)|[9, 2, 0; 9, 2, 1]; \tag{5.12}$$

In the formulation for $C_4$, the arguments in LAM indicate the direction (i.e. $U_3$) and the position of the plane of reflection (20).

The elements on the normat lines $U_3 = 0$ and $U_3 = 40$ are given by

$$C_5 = \text{LAM}(3, 20)|\text{RIN}(2, 18, 2)|[9, 0, 0; 9, 2, 0]; \tag{5.13}$$

*Transformation into cylindrical co-ordinates*

The transformation is effected by the use of the *basi-cylindrical retronorm*, which may be written as

$$r = b_1 * U_1 \tag{5.14}$$

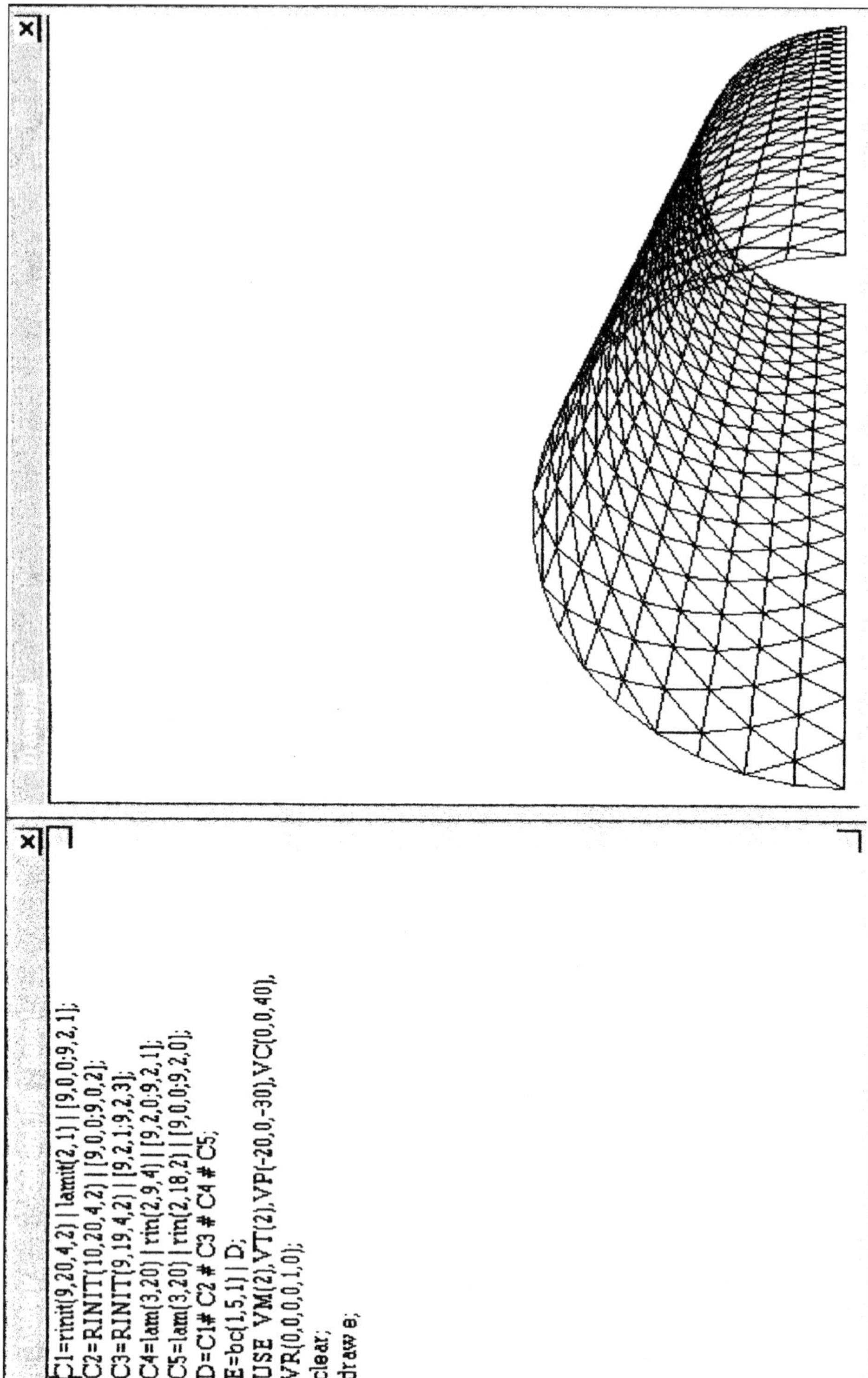


*Fig. 5.3. The perspective view of the braced barrel vault in Example 5.1*

$$s = b_2 * U_2 \tag{5.15}$$

$$z = b_3 * U_3 \tag{5.16}$$

In this case $b_1 = 1$ m and the values of $b_2$ and $b_3$ are obtained as follows.

The central angle by the semi-circular arc is 180°. Because the circumference is divided into 36 parts, the angle subtended by each subdivision is 5°. Hence $b_2 = 5$.

Because the length of 36 m is divided into 40 parts, each part is 36/40 = 0.9 m. Hence $b_3 = 0.90$.

*Formian graphics*

Formian graphics comprises concepts and constructs that enable images of Formex plots to be created on graphical output hardware such as a monitor. For example, to create a perspective view of the barrel vault on a monitor, one has to select a number of viewing parameters. These are specified by means of USE items. One must choose the view point VP($x$, $y$, $z$) and the view centre VC($x$, $y$, $z$) in terms of their Cartesian co-ordinates. The view mode VM($N$) must also be selected. $N$ takes values of 1, 2 or 3, corresponding to a *nave mode*, a *range mode* and a *zone mode*, respectively. The range mode VM(2) is selected here because it automatically positions and scales the picture. We have the choice of two view types: parallel and perspective. These are specified by choosing $N = 1$ for the parallel view and $N = 2$ for the perspective view in the USE item VT($N$). In this case we choose $N = 2$. The proper choice of VP and VC needs experience, but one can experiment with different co-ordinates for VP and VC and get a good picture. The required orientation of the picture is determined by specifying the USE item VR($x_1, y_1, z_1; x_2, y_2, z_2$). The choice of VR also needs experience. A suitable VR may be chosen by trial and error. Reference may be made to the Formian manual [5.4] for a more detailed account of Formian graphics. We are now ready to complete the Formex formulation by adding the view parameters as follows:

USE VM(2), VT(2), VP(–20, 0, –30), VC(0, 0, 40), VR(0, 0, 0; 0, 1, 0); (5.17)

The Formex formulation of the barrel vault in this example takes the following form:

$C_1$ = RINIT(9, 20, 4, 2) | LAMIT(2, 1) | [9, 0, 0; 9, 2, 1];
$C_2$ = RINIT(10, 20, 4, 2) | [9, 0, 0; 9, 0, 2];
$C_3$ = RINIT(9, 19, 4, 2) | [9, 2, 1; 9, 2, 3];
$C_4$ = LAM(3, 20) | RIN(2, 9, 4) | [9, 2, 0; 9, 2 , 1];
$C_5$ = LAM(3, 20) | RIN(2, 18, 2) | [9, 0, 0; 9, 2, 0];
$D = C_1 \# C_2 \# C_3 \# C_4 \# C_5$;
$E = bc(1, 5, 1) | D$;
USE VM(2), VT(2), VP(–20, 0, –30),VC(0, 0, 40), VR(0, 0, 0, 0, 1, 0);
clear;
draw $E$;

A print of the perspective view appearing on the monitor screen is shown in Fig. 5.3.

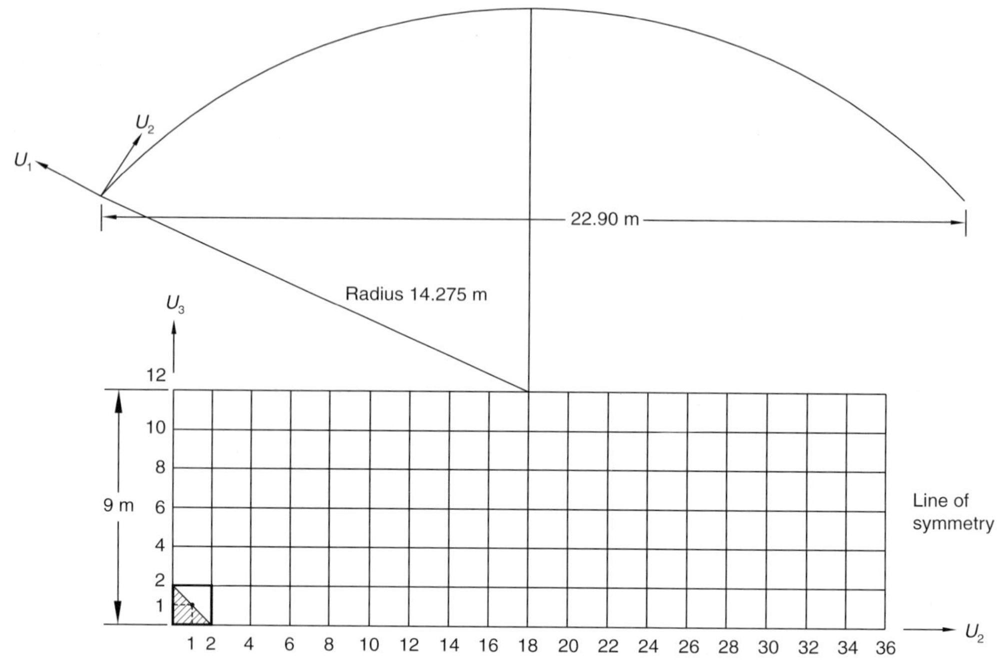


*Fig. 5.4. The intrinsic plot for the barrel in Example 5.2*

*Example 5.2*

As a second example a Formex formulation is be developed for the extrados of a barrel vault designed by the first author to cover the platform shelters of the Thirumailai Station, Chennai, India. The design of this barrel vault is described in detail in Chapter 7.

A plan and a cross-section view of this barrel are shown in Fig. 5.4. The chord width of the barrel vault is 22.90 m, the radius is 14.275 m and the angle subtended by each is 106.66°. The extrados of the vault is divided into 36 subdivisions and hence the angle subtended is 2.9628°. The barrel vault length of 9 m is divided into six divisions and hence each division is 1.5 m long. The $U_1$, $U_2$ and $U_3$ directions are chosen to lie along the radius, the circumference and the length of the barrel, respectively. It is clear that the scale factors in the basi-cylinder retronorm are $b_1 = 1$, $b_2 = 106.66/36 = 2.9628$ and $b_3 = 1.5$.

Referring to the shaded triangle in Fig. 5.4, the three lines involved may be represented by three signets:

$$\text{tri} = [14.275, 0, 0; 14.275, 2, 0; 14.275, 0, 2]; \tag{5.18}$$

The box formed by the bold lines may be represented by the equation:

$$\text{box} = \text{ROS}(2, 3, 1, 1, 2, 180) \,|\, \text{tri} \tag{5.19}$$

where ROS is a rotational Formian function known as a *rosette*. The first two arguments 2 and 3 denote the plane defined by the normat lines $U_2$ and $U_3$ about which rotation takes place; The pair 1, 1 specifies the point about which rotation takes place, 2 denotes the number of replications and 180 is the angle of rotation. In developing the box concept we have followed Lewis and Lewis [5.7].

The Formex formulation for the barrel vault in this example will now take the following form:

tri = [14.275, 0, 0; 14.275, 2, 0; 14.275, 0, 2];
box = ROS(2, 3, 1, 1, 2, 180) | tri;
*f* = LAMIT(18, 6) | RINIT(9, 3, 2, 2) | box;
*g* = *bc*(1, 106.66/36, 1.5) | *f*;
clear;
USE VT(2), VM(2), VP(20, 10, –10), VC(0, 0, 10), VR(0, 0, 0, 0, 0, 1);
*k* = LAMIT(15, 6) | [12.775, 0, 0; 12.775, 1, 0];
*l* = LAMIT(15,6) | RIN(2, 15, 2) | [12.775, 1, 0; 12.775, 3, 0];
*m* =*k* # *l*;
*n* = *bc*(1, 3, 1.5) | *m*;
*p* = *g* # *n*;
draw *p*;

A printout of the perspective view of this barrel vault is shown seen in Fig. 5.5.

## 5.7. Generating the configurations of domes

Sanchez Alvarez [5.8] has shown how dome configurations can be generated from their two-dimensional intrinsic plots. We shall illustrate this powerful technique with the aid of two examples.

### *Example 5.3*

The intrinsic plot of the dome considered in this example is shown in Fig. 5.6. The shaded triangle is represented by a Formex tri:

tri = [10, 1, 1; 10, 3, 1; 10, 3, 3]; (5.20)

By rotating the tri about point P (2, 2) through 180° we get two replications, which result in

box = ROS(2, 3, 2, 2, 2, 180) | tri (5.21)

The intrinsic plot is generated by

*f* = RINIT(16, 8, 2, 2) | box (5.22)

Conversion into spherical co-ordinates is achieved by introducing the *basi-spherical retronorm* BS($b_1$, $b_2$, $b_3$), which leads to

$$r = b_1 * U_1 \tag{5.23}$$

$$s = b_2 * U_2 \tag{5.24}$$

$$t = b_3 * U_3 \tag{5.25}$$

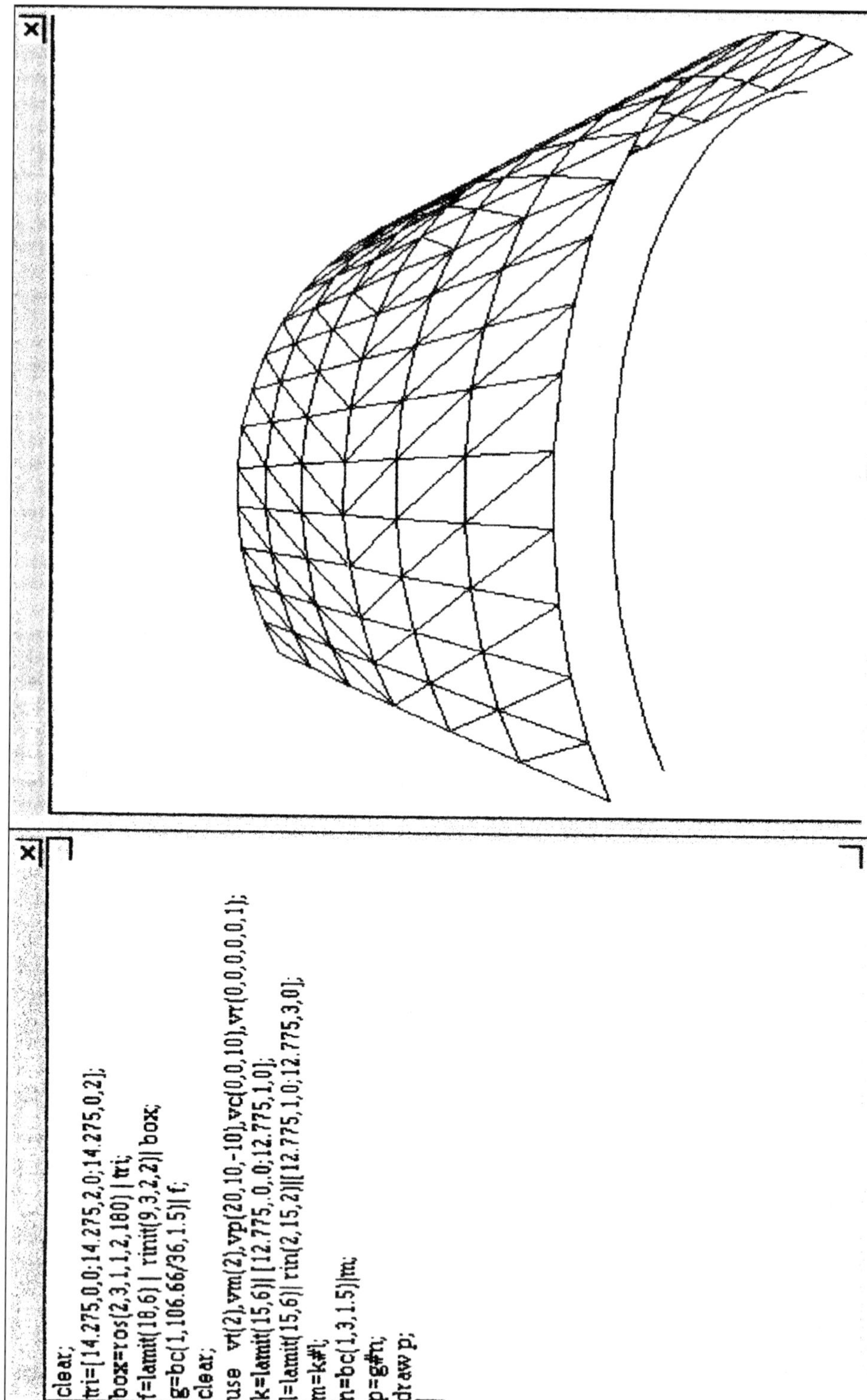


*Fig. 5.5. The perspective view of the braced barrel vault in Example 5.2*

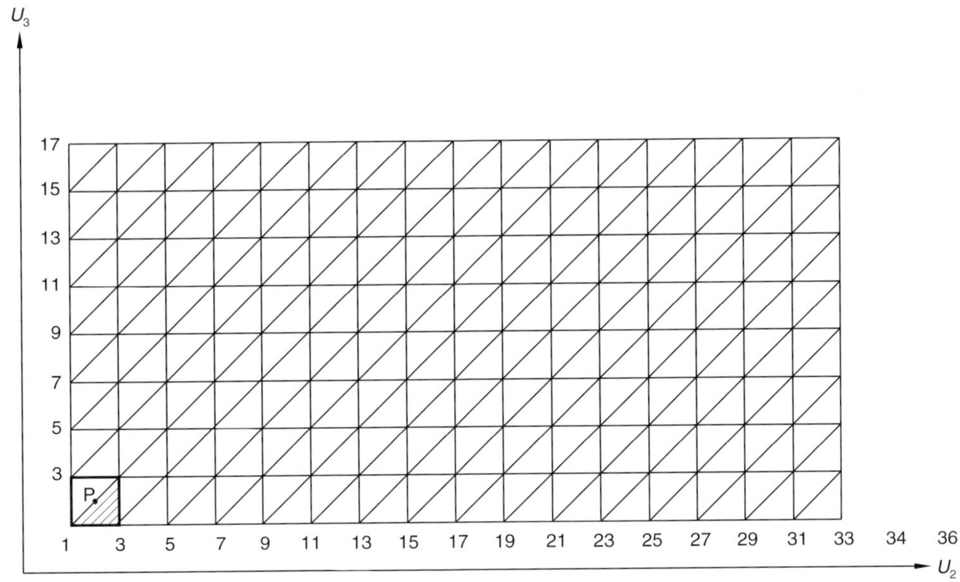


*Fig. 5.6. The intrinsic plot for the barrel in Example 5.3*

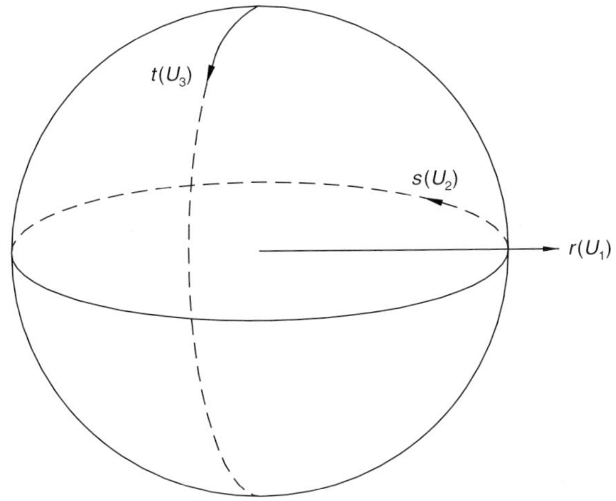


*Fig. 5.7. The basi-spherical retronorm*

The meaning of this transformation will be better understood by referring to Fig. 5.7. In this case, $b_1 = 1$, the radius $r$ is taken as 10. The value of $b_2$ is $360/32 = 11.25$ and $b_3 = 90/16 = 5.625$.

With these explanations the formulation that follows for generating the configuration of the dome considered in this example will be readily understood.

```
tri = [10, 1, 1; 10, 3, 1; 10, 3, 3];
box = ROS(2, 3, 2, 2, 2, 180)|tri;
f = RINIT(16, 8, 2, 2)|box;
g = BS(1, 360/32, 5.625)|f;
```

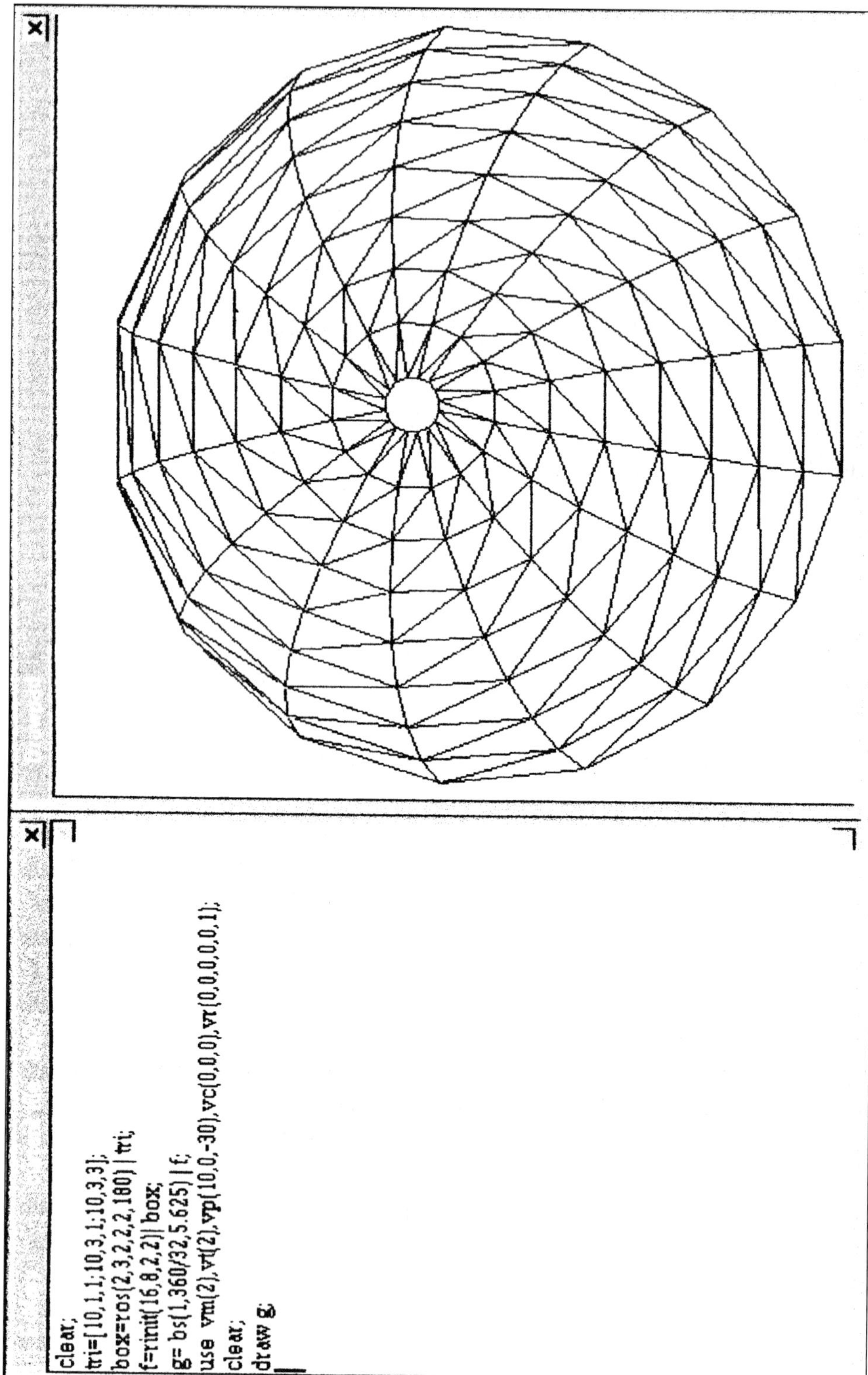


*Fig. 5.8. The perspective view of the dome in Example 5.3*

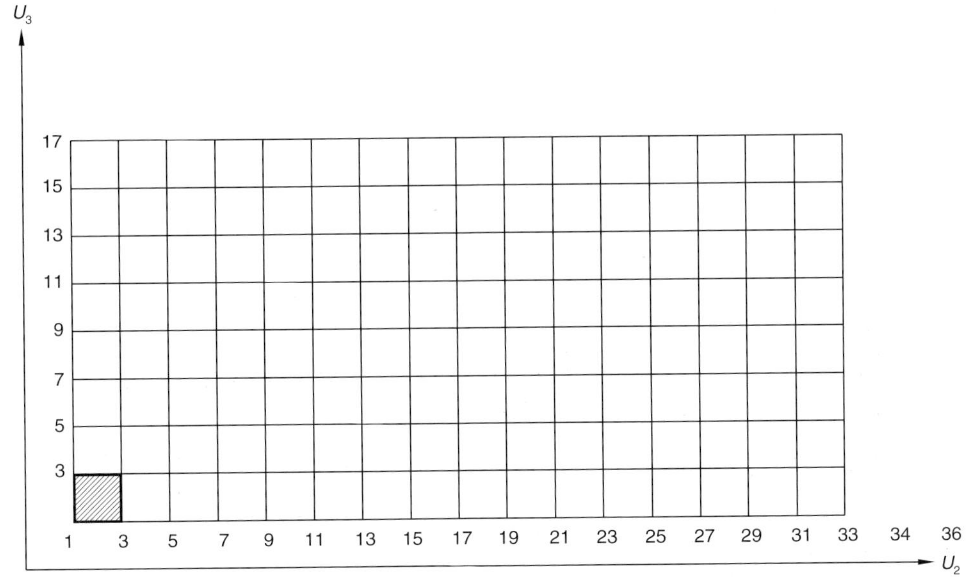


*Fig. 5.9. The intrinsic plot for the barrel in Example 5.4*

USE VM(2), VT(2), VT(10, 0, –30), VC(0, 0, 0), VR(0, 0, 0, 0, 0, 1);
clear;
draw *g*;

The perspective view of the configuration is shown in Fig. 5.8.

*Example 5.4*

The intrinsic plot of the dome considered in this example is shown in Fig. 5.9. The Formex representing the shaded square is written with four signets as:

$$\text{box} = [10, 1, 1; 10, 3, 1; 10, 3, 3; 10, 1, 3]; \tag{5.26}$$

The two-dimensional intrinsic plot is generated by:

$$a = \text{RINIT}(16, 8, 2, 2) \,|\, \text{box} \tag{5.27}$$

Spherical transformation is effected by means of the basi-spherical retronorm, with $b_1 = 1$, $b_2 = 360/32 = 11.25$ and $b_3 = 90/16 = 5.625$.

The Formex formulation for the dome in this example may then be written as follows:

*a* = RINIT(16, 8, 2, 2) | [10, 1, 1; 10, 3, 1; 10, 3, 3; 10, 1, 3];
*b* = BS(1, 360/32, 5.625) | *a*;
USE VT(2), VM(2), VP(10, 10, –40), VC(0, 0, 0), VR(0, 0, 0, 0, 0, 1);
clear;
draw *b*;

The perspective view of the dome that is generated is shown in Fig. 5.10.

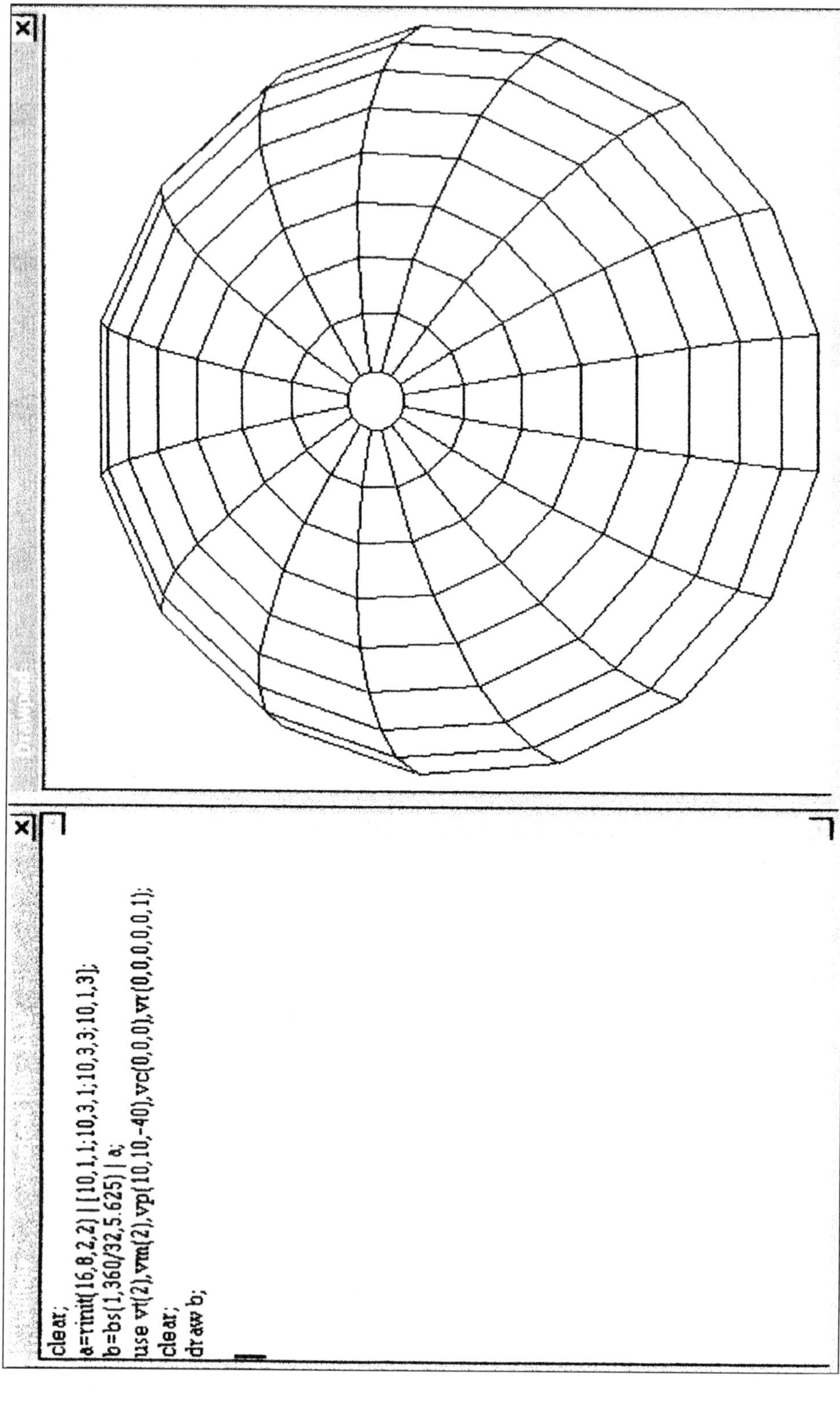


*Fig. 5.10. The perspective view of the dome in Example 5.4*

## 5.8. References

5.1. NOOSHIN, H. Formex data generation. *Proceedings of the IASS Symposium at Bangalore on Innovative Applications of Shells and Spatial Forms*, Oxford & IBH, Bangalore, 1989, vol. III, pp. 134–146.

5.2. NOOSHIN, H. *Formex Configuration Processing in Structural Engineering*, Elsevier Applied Science, London, 1984.

5.3. RAMASWAMY, G. S. and PRAKASH, N. S. Formex configuration processing for a multi-layered long span aircraft hangar. *International Journal of Space Structures*, 1999, **14**(3), 225–232.

5.4. NOOSHIN, H., DISNEY, P. and YAMAMOTO, C. *Formian*, Multi-Science, Brentwood, 1993.

5.5. DU CHÂTEAU, S. Architectural aspects of braced barrel vaults. In: Z. S. Makowski (ed.) *Analysis, Design and Construction of Braced Barrel Vaults*, Elsevier Applied Science, London, 1985, pp. 223–237.

5.6. YASSAEE, M. H. Formex formulation of the space structures of S. du Château. In: H. Nooshin (ed.) *Proceedings of the Third International Conference on Space Structures*, Elsevier Applied Science, London, 1984, pp. 117–123.

5.7. LEWIS, W. J. and LEWIS, T. S. Application of Formian and dynamic relaxation to the form-finding of minimal surfaces. *International Journal for Shell and Spatial Structures*, 1996, **37**(122), 165–186.

5.8. SANCHEZ ALVAREZ, J. S. Formex formulation of braced domes. In: Z. S. Makowski (ed.) *Analysis, Design and Construction of Braced Domes*, Granada, London, 1984, pp. 175–224.

# 6. Space trusses for long spans

## 6.1. Introduction

For spans of approximately 80 m and over, double-layered flat space frames become uneconomical. In this chapter, two examples of conceptual designs for long-span applications of space trusses are presented. The first is a triple-layer space truss for an aircraft hangar and the other is a double-layer elliptical paraboloid for covering an existing football stadium.

## 6.2. Conceptual design for an aircraft hangar for Air India at Mumbai to service 747-400 aircraft

Airlines all over the world have been using larger and larger aircraft in recent years to meet the spiralling demands of air traffic. The bulk of the investment involved in providing matching ground services is accounted for by the cost of hangarage [6.1]. In an effort to prune costs and make hangars functionally more efficient, more attention is now being paid to the following aspects of planning and design:

- choosing dimensions that optimize the area and volume required to service a particular type of aircraft
- selecting appropriate structural forms, such as space frames, and optimizing them to minimum weight
- suspending services, such as docking systems and cranes, from the roof structure itself to free floor space.

### *6.2.1. Planning the hangar layout and selecting the structural system for the hangar roof*

If only one 747 aircraft is to be serviced at a time, the minimum length and width of the hangar must be 92.5 m and 85 m, respectively. If two 747 aircraft are to be accommodated, the required span of the hangar ranges from 138 m to 154 m. These are very long spans, and a space truss is the logical choice. In connection with the design of the Heathrow hangar in the late 1960s, Makowski [6.2] made a detailed comparison of ten different structural schemes. He concluded that for very large spans the deflection limit is vital and the overall rigidity of the structure is extremely important, in particular because the roof must be capable of supporting equipment weighing nearly 600 t, of which 200 t are moveable. Another important consideration that influenced the choice is that the inherent rigidity of space frames would limit damage from accidental fires to local areas without endangering the safety of the entire structure. For these reasons, a double-layer welded diagrid space frame was chosen for the Heathrow hangar. Following the example of Heathrow, space frames have been adopted as the structural scheme for almost all hangars built in recent years to service Jumbo Jet aircraft. An important investigation undertaken for Constrado (now the Steel Construction Institute) [6.3] at the Space Structures Research Centre, University of Surrey, UK, subsequent to the construction of the Heathrow hangar, concluded that the limiting span for double-layer space trusses is in the region of 100 m and that for larger spans, because the diagonal bracing members become too long, a triple-layer space truss is the logical answer. For more detailed information on current trends in the planning, design and construction of aircraft hangars, the reader is referred to the recent paper by Ramaswamy [6.4].

### *6.2.2. Design brief for the Air India hangar at Mumbai, India*

The hangar is planned to service two 747-400 airplanes with one aircraft docked in the 'tail-in' and the other in the 'nose-in' position. The docking system and the electric overhead travelling cranes are to be suspended from the hangar roof. The hangar is to be equipped in the bay for the aircraft docked in the 'tail-in' position with a full suspended docking facility, undercarriage lifting platforms and fixed hydraulic systems. Two 10-t and two 3-t overhead cranes, compressed air, fire protection systems and other specialized ground support services for complete check-over of the planes are to be provided across the entire area covered by both bays. An annexe comprising a basement, ground floor and two upper floors with 15 000 m$^2$ of floor space, and provision for future expansion by the addition of five more storeys is to form part of the hangar complex.

### *6.2.3. Hangar dimensions*

The functional requirements detailed in the design brief were met by adopting a clear depth of 89 m and a clear width of 129.384 m for the hangar in plan (Fig. 6.1).

### *6.2.4. Choice of topology*

A triple-layer space frame, 5.53 m deep, with a configuration of diagonally laid top chords over square middle chords over diagonally laid bottom chords was selected in view of the long unsupported span of 130.82 m. The middle chords have hardly any structural function. Being located at the neutral surface, they carry only small forces. Their main function is to reduce the length of the bracing diagonals.

The five-layer front girder is formed by adding a top and bottom flange to the triple-layer space frame. However, unlike the rest of the space frame, the front girder is of welded construction. The top and bottom flanges, comprising longitudinal and cross girders are closed box structural sections, as are the diagonal bracings connecting these flanges to the top and bottom chords of the triple-layer

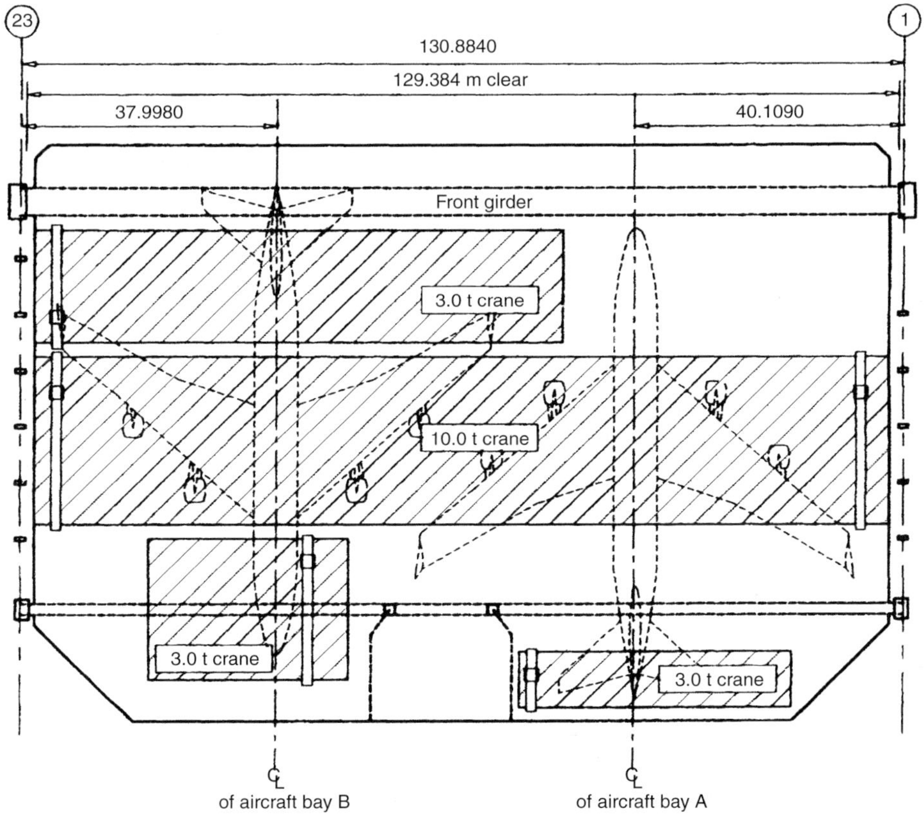


*Fig. 6.1. Plan of the aircraft hangar, Air India, Mumbai, India. (Reproduced with permission from Joglekar* et al. *[6.5])*

space frame. The depth of the front girder is 9.53 m. The top-most and bottom-most layers of the five-layer front girder are referred to here as 'flanges'. A cross-section of the hangar is shown in Fig. 6.2. The layout of the top and bottom flanges, the top, bottom and middle chords of the space frame and the bracing diagonals is contained in the CD-ROM that accompanies this book. The space frame is supported on the nodes of the middle chord.

### 6.2.5. *Structural analysis and design*

The structural analysis was done using the SAP90 software. In the preliminary analysis it was assumed that the space truss was pin jointed. In the final analysis the front girder was regarded as rigid. The sizes of the members chosen for the preliminary run and the support conditions assumed are contained on the accompanying CD-ROM. Taking advantage of the symmetry about the *y* axis, only half of the structure was analysed. The computer output of the forces and displacements are on the CD-ROM. The maximum deflection under a working load of 624 mm was found to be at node 1424.

### 6.2.6. *From conceptual to final design*

The conceptual design for the hangar developed by the first author formed the basis for the final design, which incorporated the following changes:

- the structural depth of the space frame was reduced to 4.871 m
- the front girder depth was restricted to 8.683 m
- a rear girder was introduced to rest on two already available columns
- the square configurations of the top and bottom flanges were replaced by diagonal layouts
- the front girder was pre-cambered to 500 mm.

In the final design, the triple-layer space frame was regarded as pin-jointed and the front girder as rigid jointed. Loads considered included dead, live, wind and seismic loads. The Indian Code of Practice (IS 875: Part 3) requires a dynamic analysis for wind for the main structural components of any structure with the following properties:

- height to main plan dimension greater than 5
- a first mode of vibration less than 1 Hz.

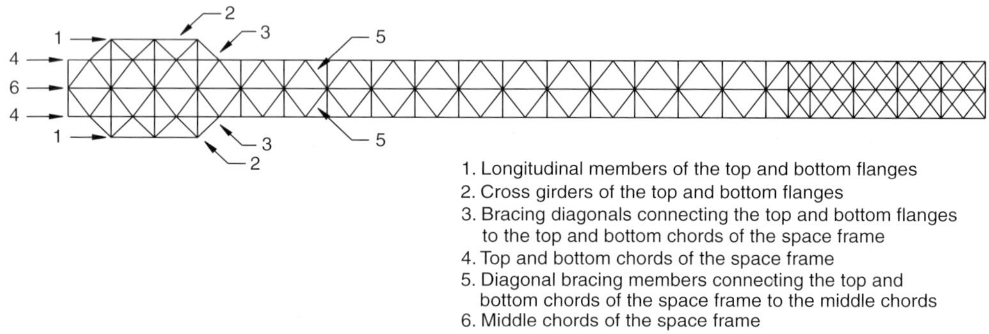


*Fig. 6.2. Cross-section of the hangar*

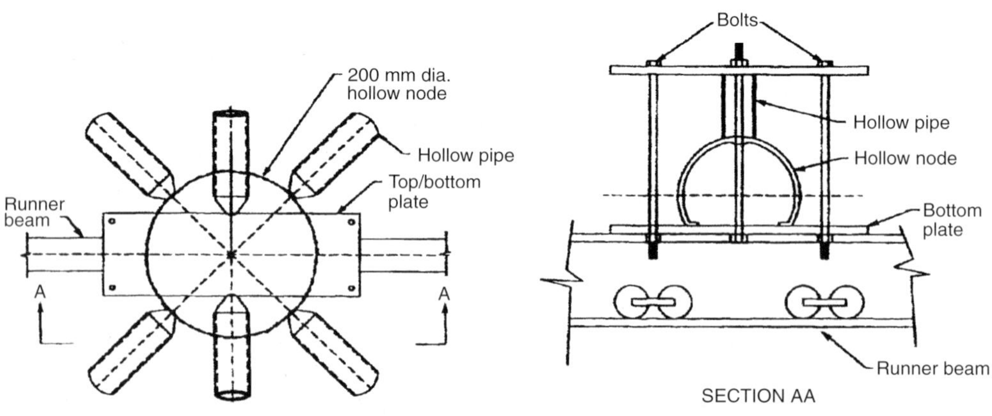


*Fig. 6.3. The docking system. (Reproduced with permission from Joglekar* et al. *[6.5])*

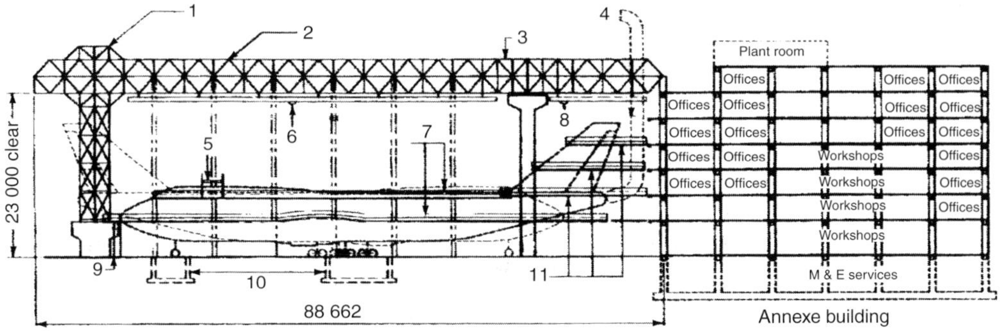


1. Front girder
2. Triple-layer space frame
3. Rear girder
4. APU exhaust
5. Mobile suspended fuselage platform
6. 10-t travelling crane
7. Upper and lower suspended fuselage docks
8. 3-t travelling crane
9. Mobile nose platform
10. Undercarriage pits
11. Suspended tail docks

*Fig. 6.4. Cross-section of the hangar after completion. (Reproduced with permission from Joglekar* et al. *[6.5])*

As the computed natural frequency of the hangar was 1.513 Hz, no dynamic analysis for wind was called for. For the same reason, no wind-tunnel tests were considered necessary.

### *6.2.7. Roof cladding*

The space truss is clad with 0.9 mm thick galvanized steel, trapezoidal, powder-coated sheets, the thickness of the coating being 40 μm.

### *6.2.8. Docking system*

The docking system design was by Clyde Carruthers, a division of Clyde Industries Ltd. It involves tail, fuselage and nose docks, which operate on runner beams. The beams themselves are suspended from the space frame (Fig. 6.3).

### *6.2.9. Node connectors*

Hollow Tuball node connectors of the Octatube system were used for the top and bottom chords. A solid node was used to connect members at joints in the middle chord. In the front and rear girders the maximum forces in the top and bottom flanges were 1500 t tensile and 1600 t compressive. To transmit these forces and to facilitate connection with the members, a cylindrical joint with fin plates radially welded to it was developed.

### *6.2.10. Hoisting and erection*

The hangar, weighing 1100 t, was assembled on the ground and lifted in one piece using centre-hole PSC jacks. A cross-section of the hangar as finally built is shown in Fig. 6.4. For a more detailed account of this project, the reader is referred to the paper by Joglekar *et al.* [6.5].

### *6.2.11. Credits*

The overall planning, including that for equipment and services, was entrusted to M/s. Scott Wilson Kirkpatric, UK, by Air India, who are the owners. The final design was evolved in the office of STUP Consultants Ltd, Mumbai, India, based on the conceptual design developed by the first author. IRCON International acted as management consultants and the contractors were OCTAMEC Engineering Ltd.

## 6.3. Conceptual design of a double-layer elliptical paraboloid for the total roof enclosure of the Sydney Football Stadium, Australia

Because of the difficulty experienced in growing a wearable natural grass in the Astrodome at Houston in 1960, American football is played in indoor stadia on synthetic grass. However, soccer and rugby football codes prohibit the use of artificial grass surfaces on the grounds that they are too hazardous for the players. Hence these games continue to be played in outdoor stadia. Outdoor stadia have several disadvantages:

- they do not provide adequate protection for spectators from exposure to the sun, rain and cold
- they generate only limited revenue because they are unsuitable for multiple event usage.

Indoor grass surfaces are not considered tough enough for vigorous play and, even if the other horticultural problems could be solved, there is still the problem of the prohibitively high cost of removing the heat in order to maintain a suitable temperature for healthy grass growth.

### *6.3.1. The Sydney Football Stadium*

The Sydney Football Stadium was opened in 1986. It has 42 000 seats, of which approximately 20 000 are covered. The stadium is used for rugby and soccer, and a few additional events.

Following several rain-affected events and consequent community and media criticism, the owner of the stadium, The Sydney Cricket and Football Ground Trust, commissioned an Australian group with a background in the design and construction of lightweight roofs to report on the technical feasibility of extending the existing roof to provide cover over all the seats. The mandate given to the group was to come up with a solution that would:

- resolve the long-standing problem of cultivating a wearable, shade-tolerant, indoor grass for the playing surface, and at the same time remove the heat economically
- permit the game to be played all the year round and make such stadia multi-purpose facilities in order to generate the extra revenue to make them financially viable
- shelter spectators from the rain, cold and sun
- reduce shadows, in order to permit television coverage

- use materials optimally in order to achieve a minimum-weight roofing scheme
- house services within the structure, while it is being assembled on the ground, in order to eliminate the expense and hazards of working at heights.

The curator of the stadium, who was co-opted as a member of the team, calculated the quantity of light required by the grass surface, assuming that there would be a change to shade-tolerant grass, which was under development elsewhere. The conclusion reached was that for such grass to grow there needed to be at least 70% light transmission through the extended roof.

### *6.3.2. Structural schemes for the roof*

Among the structural schemes explored were the following:

- A retractable roof, which was ruled out as a possibility because it was prohibitively expensive.
- A cable supported design was considered, but rejected because the fabric to be used provided no more than 30% light transmission. Other factors involved in reaching this decision were the unacceptably low tensile and tear strength of the material, which made it unsuitable for use as a tensioned membrane. These fabrics also blocked most of the UV-A wavelengths required for strong leaf growth and colour.

### *6.3.3. Roof cladding*

The solution for the roof cladding was found in an unreinforced material that had adequate durability and light-transmission properties. The fabric cladding proposed was to be contained in rectangular aluminium panels. It is fastened to the steel support layer over the structural frame. Each panel contains two layers of film separated by a constant supply of low-pressure air, and thus acts as a pillow. The double-layer pillows provide insulation benefits, and the convex internal surfaces serve to offset the refocusing of sound caused by the global concave profile of the roof. The damping caused by the air gap and the selection of different film thicknesses for the external and internal layers further improves the acoustics. A secondary partial ceiling layer, composed of a fabric of appropriate thermal and acoustical properties, beneath the roof structure was proposed in order to facilitate the control of noise and temperature.

### *6.3.4. Meetings at Delft, The Netherlands*

The Australian team, consisting of Mr I. Norrie and Mr Thorogood of Stadium Technologies, Sydney, arrived at Delft in August 1991 to meet Prof. Dr Eekhout, Director General of Octatube BV, The Netherlands, and Professor Ramaswamy, Chairman, Octatube Space Structures (India) Pvt. Ltd. The object of the meeting was to give final shape to the double-layer space frame proposal in consultation with experts. The Australian team had already resolved the problem of the roof cladding. They had also arrived at the tentative conclusion that a double-layer tubular space frame supported on four columns and 2 m above the existing roof offered the best solution for enclosing the stadium (Fig. 6.5). Exhaust fans were to be installed in the gap between the structures. This passive energy method was chosen because it provides adequate spectator and player comfort at an affordable cost.

### 6.3.5. Geometry of the space frame

The plan of the proposed space frame is an ellipse, with a semi-major axis of 92.5 m, a semi-minor axis of 74 m (Fig. 6.6) and a rise of 25 m. Choosing the origin of the co-ordinates to lie 25 m below the apex, and aligning the $x$ and $y$ axes along the semi-major and semi-minor axes, respectively, and $z$ axis directed vertically upwards, the equation for the top surface of the space frame may be written as:

$$z = 25 - 0.00292184\,x^2 - 0.001022644\,y^2 \qquad (6.1)$$

The shape of the surface is shown in Fig. 6.7.

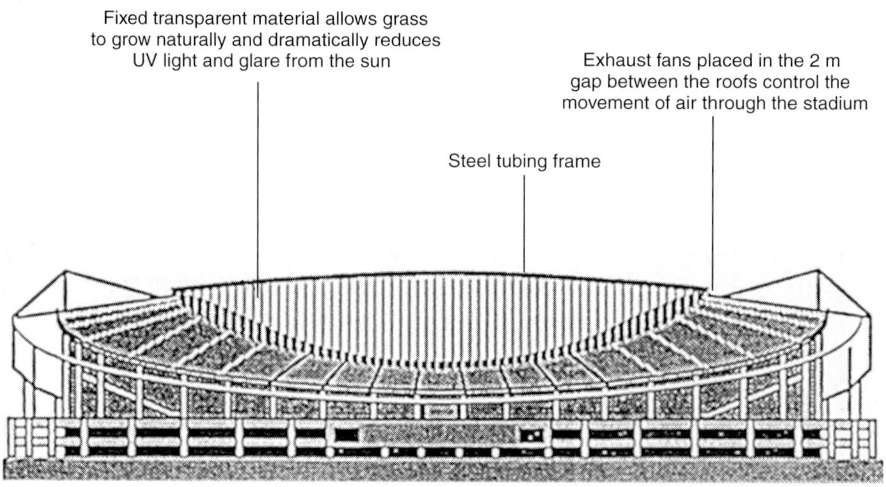


*Fig. 6.5. The proposed space frame for enclosing the Sydney Football Stadium, Australia*

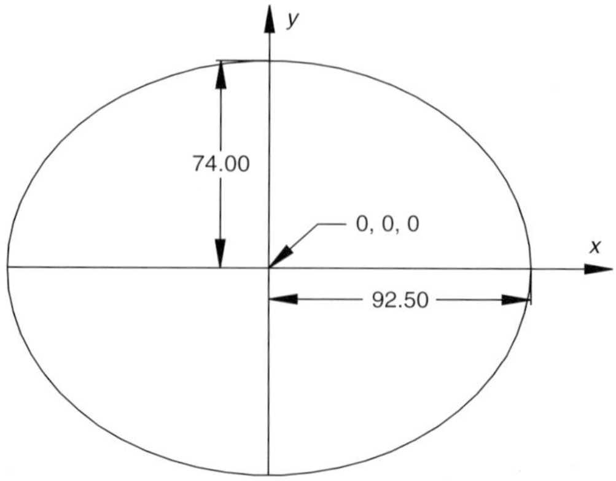


*Fig. 6.6. Plan of the stadium*

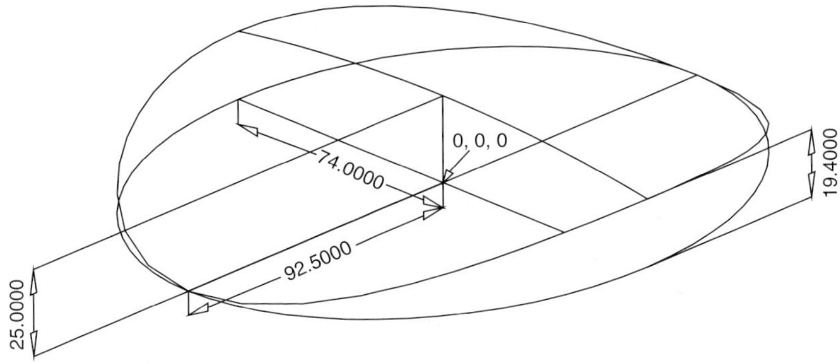


*Fig. 6.7. View of the stadium*

### 6.3.6. Selection of the grid topology

At the Octatube office at Delft, a stress analysis was performed for space frames with ten different combinations of topology and structural depth. After a comparison of their relative merits, a square over square offset configuration diagonally set with a structural depth of 5 m and a 5 m × 5 m diagonal grid was selected. The choice was governed by the need to control deflections.

The space truss has no ring beam, but the edges are stiffened with several tubes of very large diameter. A narrow band of heavy tubes forming a cross connects the four columns. It is across this band that the load transfer to the columns takes place. Each column is a suitably braced tripod. The location of the columns is shown in Fig. 6.8, and a summary of the design loads is given in Table 6.1. For purposes of stress analysis the space truss was regarded as pin-jointed.

### 6.3.7. Passive energy design features

One of the unique features of the proposed stadium is the extent to which passive energy features have been used for environmental control in order to minimize energy costs. The twin objectives were to ensure spectator and player comfort and to provide an environment conducive for grass-growth management when the stadium is not in use for events. The scheme in outlined in Fig. 6.9.

The estimated 800 t of air in the stadium will be changed by entirely passive means at rates ranging from 26 minutes in winter with still air, to 7.7 minutes in summer with 20 km/h external wind. Supplementary fans with a capacity of 7.8 minutes for air change are provided for smoke exhaust purposes.

### 6.3.8. Acceptance by owners

The above-described proposals have been accepted by the Sydney Cricket and Football Ground Trust and their implementation is awaiting the availability of funds. If and when completed, the totally enclosed football stadium will vie for attention with well-known Sydney landmarks such as the Central Point Tower, the Opera House and the Harbour Bridge.

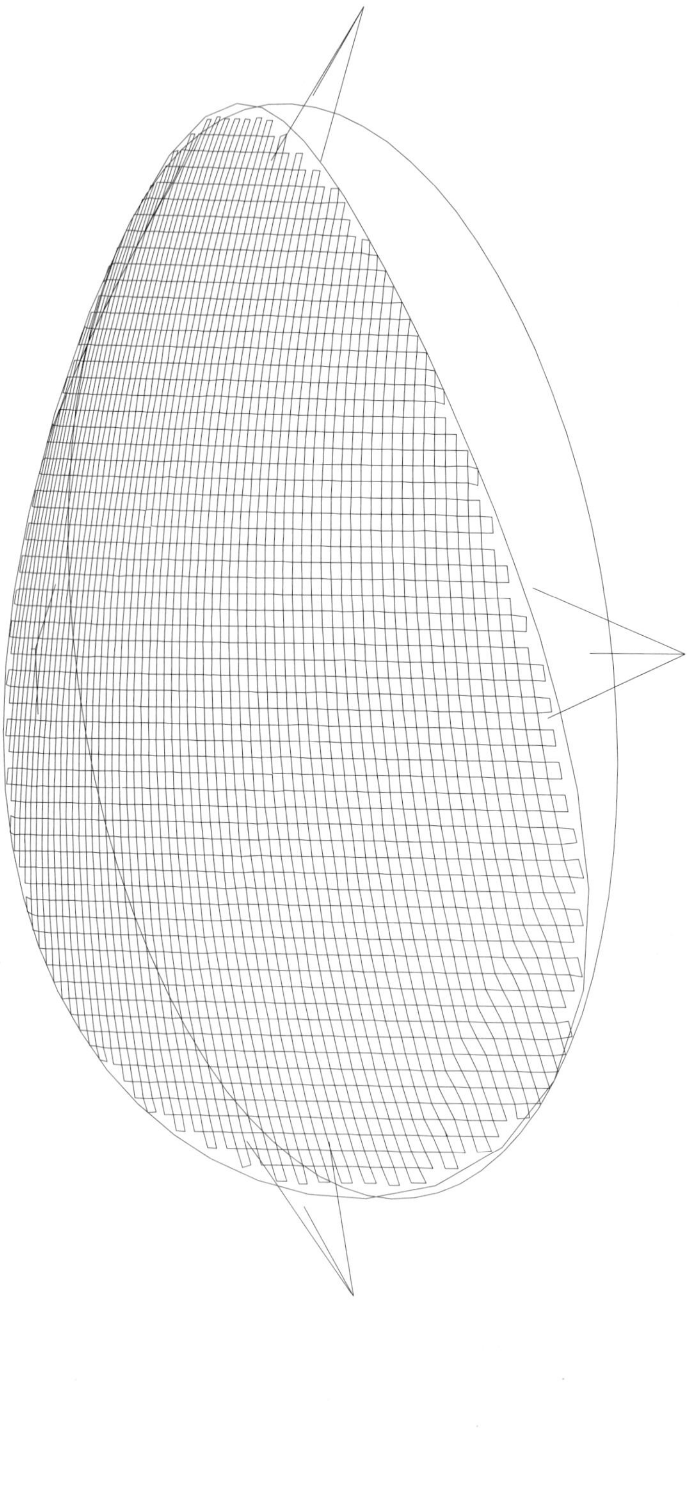

*Fig. 6.8. Perspective view of the stadium showing the locations of the columns*

*Table 6.1. Design loads for the Sydney Footbal Stadium*[a]

| Load type | Load (mass/m$^2$) | Load (t) |
|---|---|---|
| Dead load | | |
| Space frame self-weight | 65 kg/m$^2$ | 1415 |
| Nodes, additional | 3140 @ 35 kg each | 110 |
| Fabric support framing | 8 kg/m$^2$ | 175 |
| Fabric membrane, Al, clamps | 2 kg/m$^2$ | 43 |
| Sunscreen mechanical layer | 15 kg/m$^2$ | 325 |
| Catwalks, lights, loudspeakers | 5 kg/m$^2$ | 110 |
| Hanging exhibits | 16 @ 5 t each | 80 |
| Hanging scoreboard centre | 4 @ 25 t each | 100 |
| Oprable wall down centre | 30 kg/m$^2$ | 150 |
| Perimeter louvres | 50 kg/m run | 25 |
| Perimeter edge deflector | 50 kg/m run | 25 |
| Perimeter exhaust vents | 28 @ 1 t each | 28 |
| Corner columns | 4 @ 74 t | 296 |
| Total dead load | | 2882 |
| Live load | 25 kg/m$^2$ | 543 |
| Total dead and live load | | 3425 |
| Wind load | 0.98 (coeff.) kPa | 2345 |

[a]Reproduced with permission from Ramaswamy *et al.* [6.6]

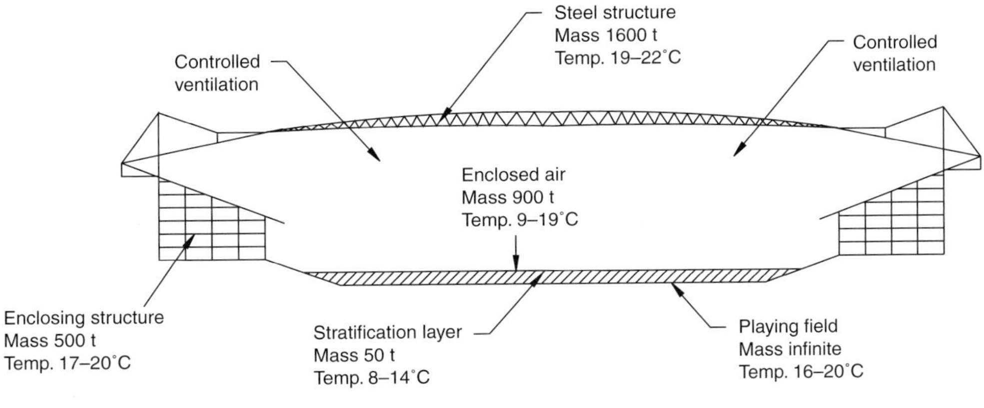


*Fig. 6.9. The passive energy design features of the stadium. (Reproduced with permission from Ramaswamy* et al. *[6.6])*

For a more detailed account of the proposals for totally enclosing the Sydney Football Stadium, the reader is referred to the paper by Ramaswamy *et al.* [6.6]. A pictorial view of the space frame is given in Fig. 6.10.

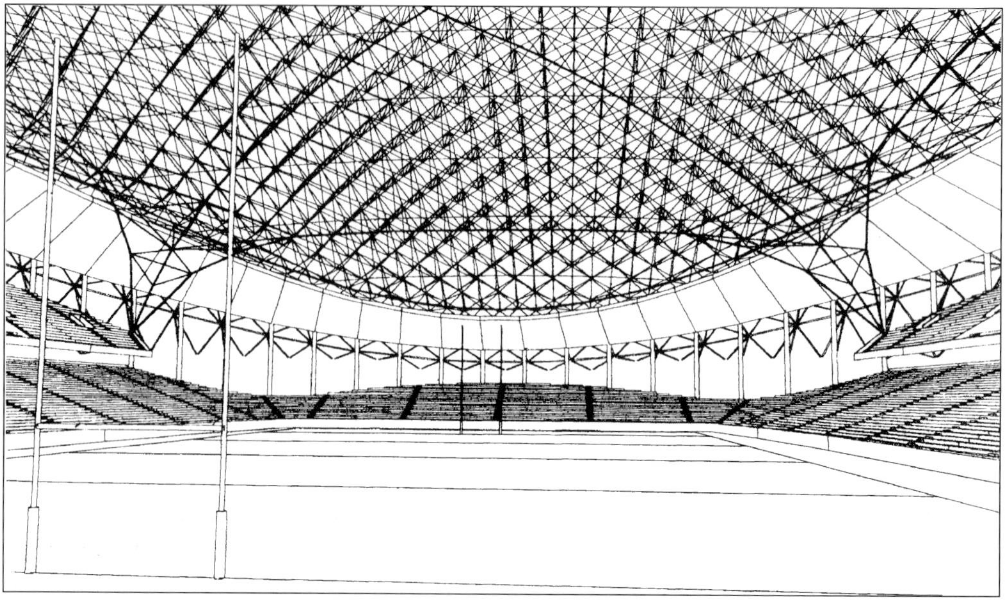

*Fig. 6.10. A pictorial view of the space frame for the stadium. (Reproduced with permission from Ramaswamy* et al. *[6.6])*

## 6.4. References


6.1. JOYNER, K. J. The provision of hangarage facilities. *Tubular Structures*, 1990, **15**, 2–4.

6.2. MAKOWSKI, Z. S. Feasibility study, design and analysis. *Tubular Structures*, 1990, **15**, 23–32.

6.3. BINNY, U. K., DISNEY, P. and MAKOWSKI, Z. S. *Multi-layer Space Frames*. A report on the investigation into triple-layer grids, carried out for CONSTRADO, at the Space Structures Research Centre, University of Surrey, Guildford. CONSTRADO, London, 1980.

6.4. RAMASWAMY, G. S. Review of recent trends in the planning, analysis, design and construction of space frame roofs for aircraft hangars. *International Journal of Space Structures*, 1999, **14**(3), 159–165.

6.5. JOGLEKAR, S. G., KHANRA, K. K, NAVANI, K. P. and DADLANI, N. S. Long span space frame hangar roof with suspended docking system and cranes for heavy maintenance of 747-400 aircraft for Air India at Mumbai, India. *International Journal of Space Structures*, 1999, **14**(3), 197–209.

6.6. RAMASWAMY, G. S., THOMAS, D. S., NORRIE, I. and THOROGOOD, L. A design for a football stadium roof enabling growth of a natural grass playing surface: Sydney Football Stadium. *Bulletin of the International Association for Shell and Spatial Structures*, 1994, **35**(2), 85–92.

# 7. Braced barrel vaults

## 7.1. Introduction

The efforts of man to build and bridge in ancient and mediaeval times were severely handicapped by the non-availability of materials with adequate tensile strength. Undaunted by this formidable obstacle, he devised structural forms, such as the arch, the vault and the dome, to suit the properties of the available materials (stone, timber and brick) that had reasonable compressive strength. The masonry vault is thus the forerunner of the modern steel braced barrel vault. The evolution of the braced barrel vault has been documented in detail by Makowski [7.1].

## 7.2. Geometry of braced barrel vaults

Braced barrel vaults are developable surfaces of zero Gauss curvature generated by the movement of a curve, known as the 'directrix', over a generator straight line. The directrix may be a circular arc, an ellipse, a catenary, a parabola or a cycloid [7.2]. Most of the barrel vaults built in practice are parts of a right circular cylinder.

## 7.3. Choice of configuration

The many different schemes of diagonal bracing that are possible result in a wide variety of configurations to choose from. These configurations may be broadly categorized as triangular and rectangular. Plans of some of the configurations that are often chosen are given in Fig. 7.1. Early investigations carried out by Makowski [7.3] on small-scale models of single-layer barrel vaults led to the conclusion that the three-way triangular configuration developed a more or less uniform stress distribution under uniformly distributed load. As a consequence, a barrel vault of this configuration can be built of members of identical size. In practice, it may be necessary to use tubes of the same outside diameter but with two or three different thicknesses. Moreover, single-layer barrel vaults with a three-way triangular configuration deflect much less under non-symmetrical loads than do other types. Studies reported by Parke [7.4] have shown that the most economical ratio of rise to chord width of cylindrical barrel vaults in relation to weight is 0.17.

## 7.4. Single- and double-layer vaults

Walker [7.5] claims that, depending on the rise/span ratio and loading, single-layer braced barrel vaults built using Nodus connectors offer a cost-effective and practical solution for transverse spans of up to 40 m. However, noting that single-layer barrel vaults are vulnerable to snap-through buckling, for transverse spans in excess of 20 m it is desirable to have some measure of rigidity introduced by welding the node connectors, as in the SDC joints employed by du Château for his design of a tennis court for SNCF at Paris–Vaugirard [7.6]. A perspective view of the structural scheme employed for this elegant barrel vault is given in Fig. 5.10 (see Chapter 5). The recently introduced Mero Plus ZK nodes with bending stiffness provided by two pre-stressed bolts are also suitable for such applications.

## 7.5. Structural analysis and design

The details of the structural analysis and design of a typical cylindrical barrel vault for roofing a railway platform shelter are presented in the case study that follows.

## 7.6. Case study: a cylindrical braced barrel vault for roofing platform shelters at the Thirumailai Railway Station, Chennai, India

The barrel vaults cover the up and down platforms of the suburban Thirumailai station of the Metropolitan Rapid Transit System for the city of Chennai, India. The fixtures for electric traction are hung from the double-layer arches spaced 9 m apart

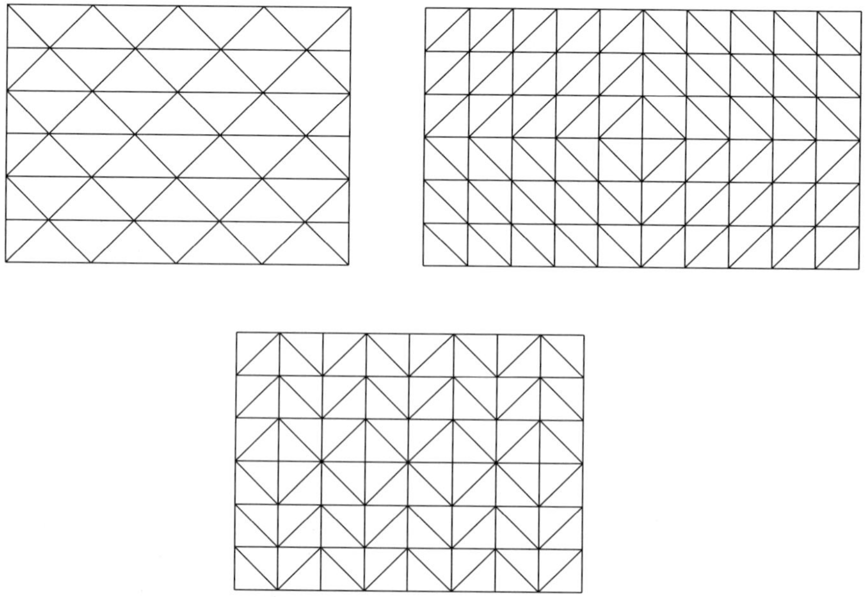

*Fig. 7.1. Plans of barrel vault configurations*

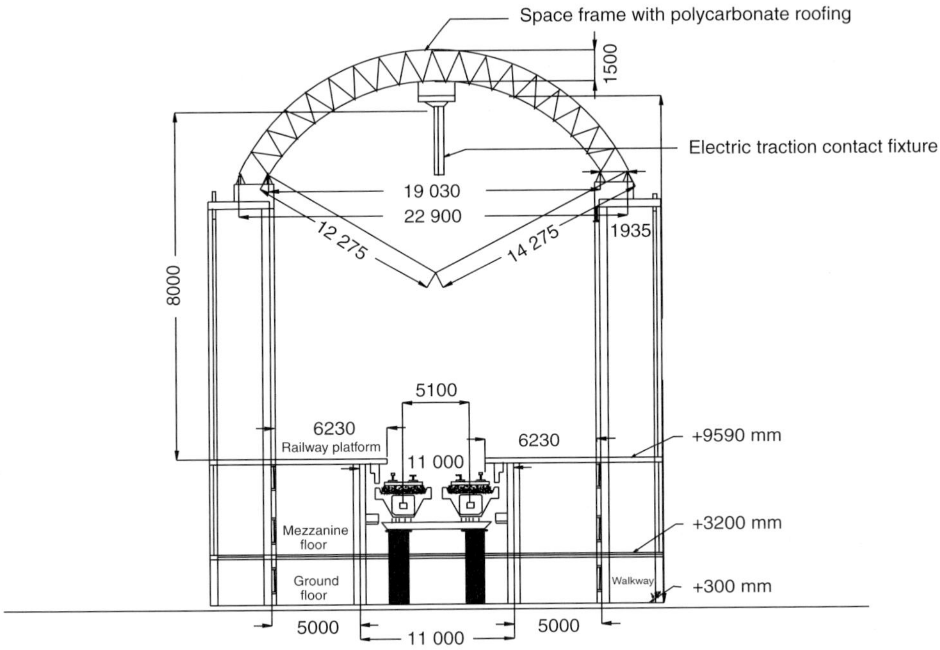


*Fig. 7.2. Cross-section of the parallel vault and railway tracks*

on which the single-layer barrel vaults are supported. The barrel vaults involved are continuous over two, three or four spans. The platforms, which are 226 m long, are aligned along a gentle horizontal curve. While heightening the aesthetic appeal of the barrels, this curve made their design somewhat more complex. A cross-section of the barrel vaults and railway tracks is shown in Fig. 7.2.

To make barrel vaults as structurally efficient as possible, it is necessary to ensure that the loads on them are transferred to supports by a combination of *beam* and *arch action*. To this end, the barrel vaults are supported on double-layer stiffening arches provided at 9 m intervals. In addition, they are rigidly welded to edge beams fabricated out of rectangular hollow sections (Fig. 7.3). The arches and edge beams also serve another important structural function – as single-layer barrel vaults are prone to snap-through buckling, the arches and edge beams help to preserve the initial geometry. The roofs are clad with polycarbonate sheets.

### 7.6.1. Geometry

The geometry of the barrel vault and the arch intrados is shown in Fig. 7.4. The intrados and extrados of the arch are arcs of concentric circles. The barrel vault arc is divided into 18 equal parts. In Chapter 5, the geometry of this barrel vault was generated by means of the Formex formulation and the resulting perspective view may be seen in Fig. 5.5.

### 7.6.2. Structural analysis

In this case study, a barrel vault continuous over three spans has been chosen to illustrate the analysis and design procedures. The plan of the barrel vault and the elevation of the stiffening arches, with node and member numbering, are given in Figs 7.5 to 7.7, respectively.

For the purposes of the present analysis, the space frame, the members of which are connected together with Tuball nodes, is regarded as pin jointed. But the edge beams are regarded as rigid jointed. Structural analysis was carried out using the SAP90 software for loads corresponding to the ultimate limit state. However, the analysis itself was linear and elastic. The structural design of the members was carried out to conform to Indian Standard IS 800: 1984. As pointed out in Chapter 2, the design of tension and compression members using IS 800: 1984 or Eurocode 3: 1989 will lead to the same member sizes.

*Loads*

Dead load:

Weight of sheeting 10 kg/m$^2$
Estimated weight of space frame, purlins, nodes and stubs 25 kg/m$^2$
Total dead load = 10 + 25 = 35 kg/m$^2$

Live load 60 kg/m$^2$

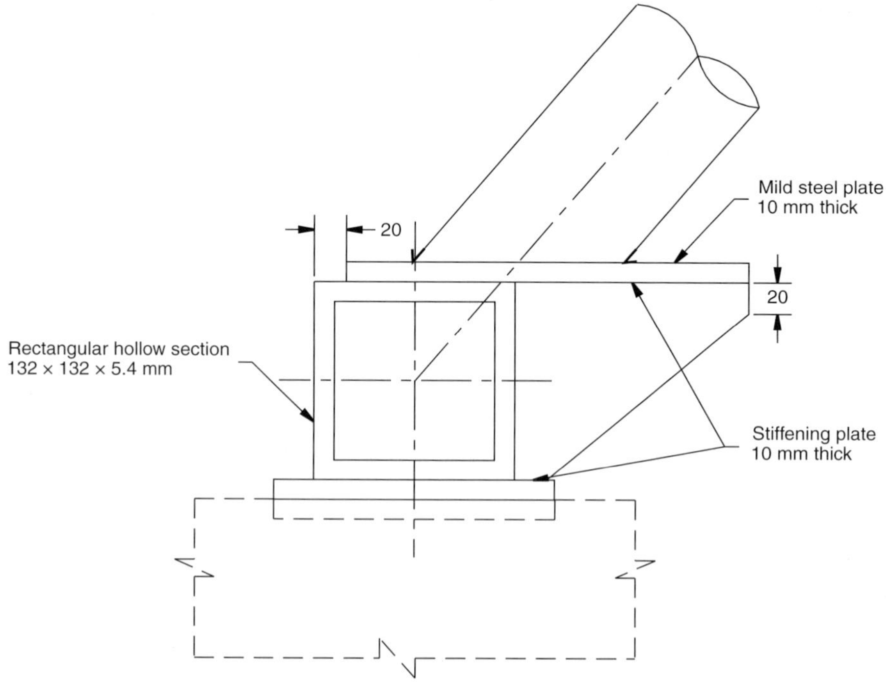


*Fig. 7.3. Detail of the connection of the edge beam and the barrel vault*

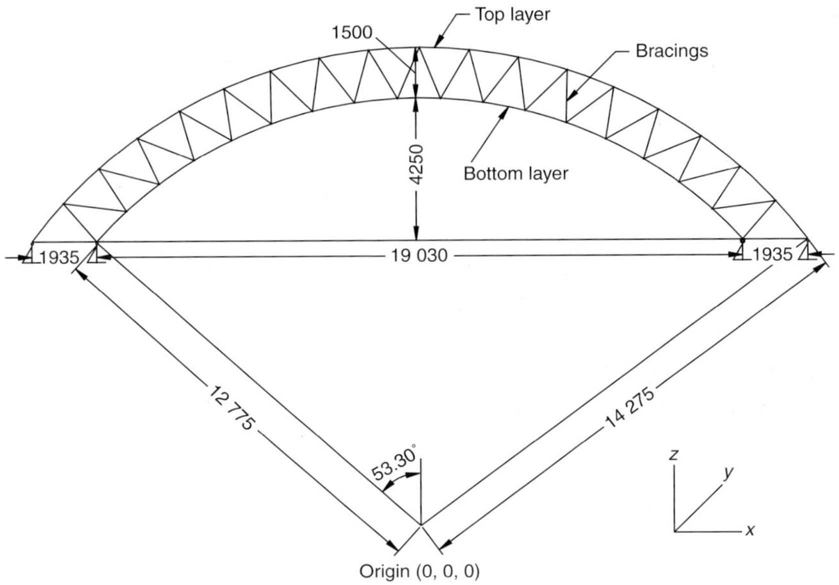


*Fig. 7.4. Cross-section of the space frame over a stiffening arch*

*Governing load combinations*

The client had prescribed that the structure needed to be analysed for a positive wind pressure of 160 kg/m² and a negative wind pressure of 240 kg/m². The two critical load combinations for which the analysis is made are as follows:

Case 1:

$$1.5(\mathrm{LD} + \mathrm{WL}) = 1.5(160 + 35) = 292.5\ \mathrm{kg/m^2} = 2.87\ \mathrm{kN/m^2} \downarrow$$

Case 2:

$$1.5(\mathrm{DL} + \mathrm{WL}) = 1.5(240 - 35) = 307.5\ \mathrm{kg/m^2} = 3.00\ \mathrm{kN/m^2} \uparrow$$

*Node and member numbering*

The node and member numbering schemes are given in Figs 7.5 and 7.6.

*Preliminary run*

The following preliminary sizes were chosen for the members of the space frame to start the analysis.

Longitudinal and transverse members of the space frame: 114.30 mm × 4.50 mm
Bracing diagonals: 60.30 mm × 4.50 mm
Intrados of arch: 60.30 mm × 4.50 mm

These sizes were used for the preliminary run using the SAP90 software. The computer output for the forces was scanned and revised member sizes were chosen to match the member forces. These data provide the input data for the next run. The weight of the structure is computed after each run. The process of iteration is continued until the difference in weights between two successive runs is negligible. Three runs were required in this instance. The weight converged to 8.25 t and the corresponding steel consumption was calculated as 11.5 kg/m$^2$. The computer input data and the output in the form of forces, displacements and reactions for the last run are contained on the accompanying CD-ROM.

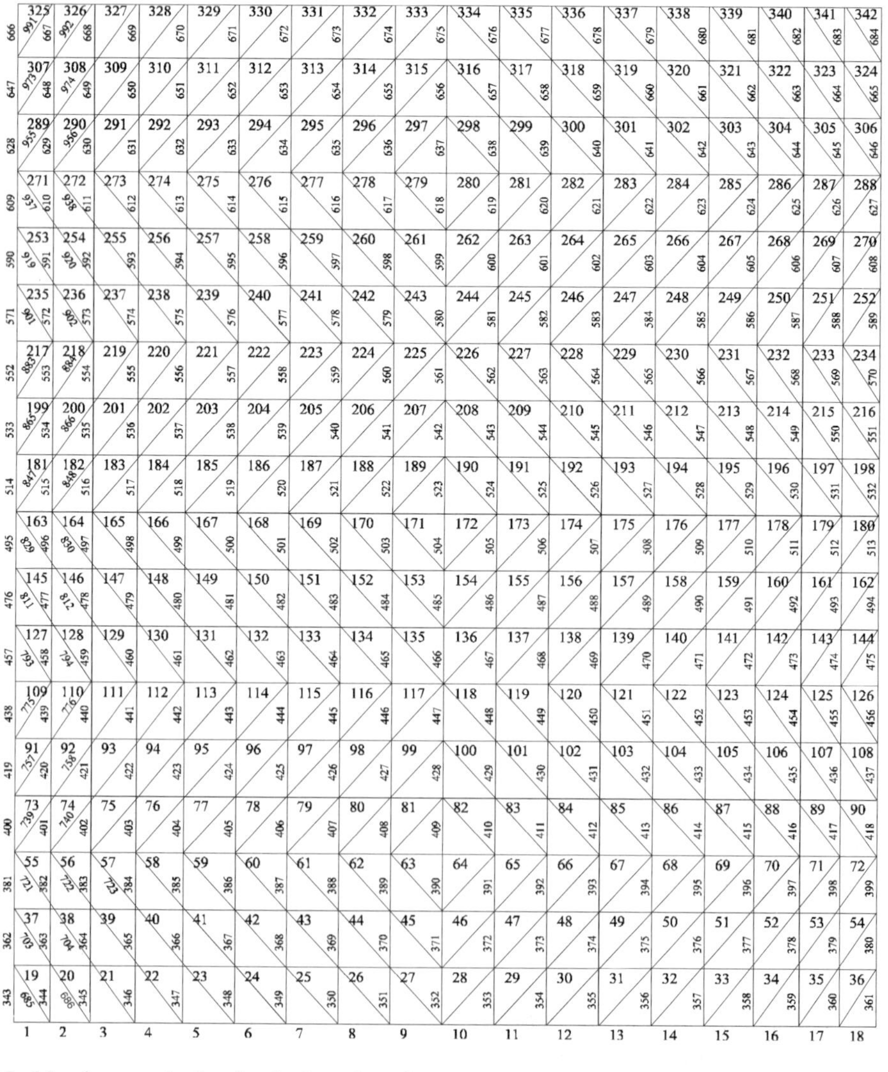

*Fig. 7.5. Member numbering for the barrel vault*

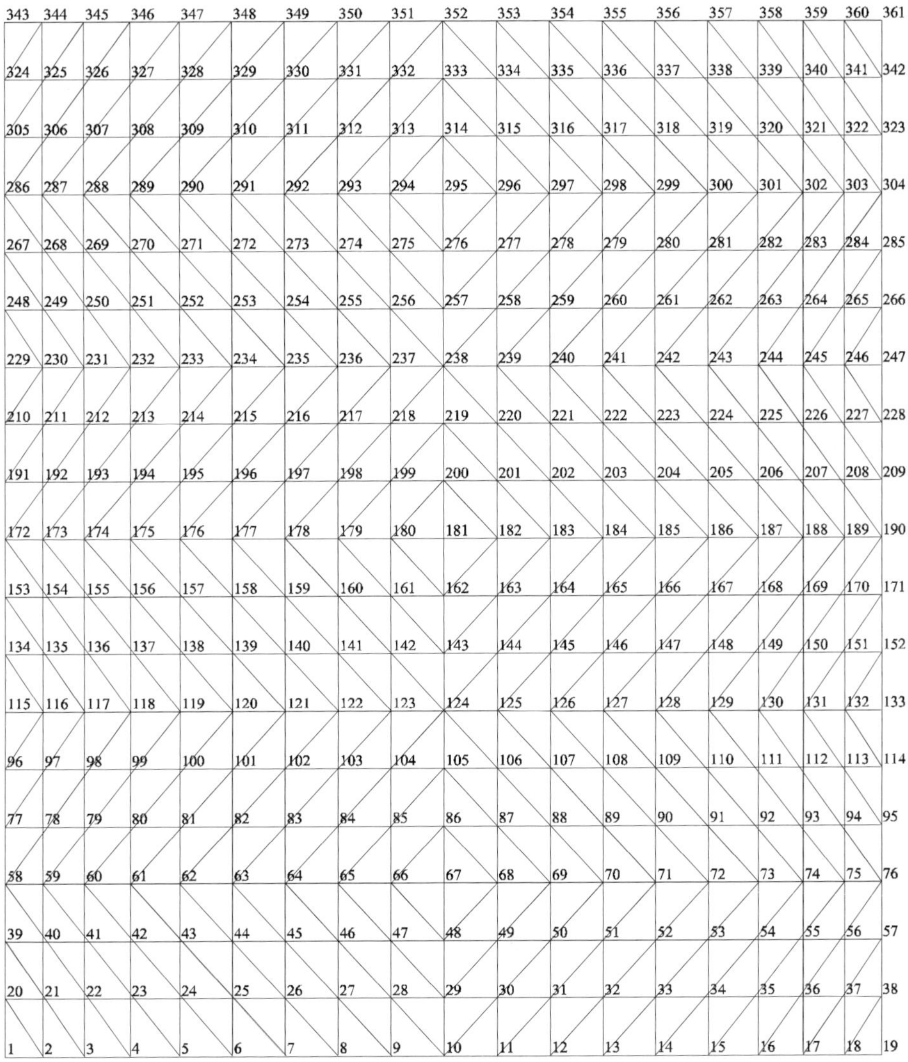


*Fig. 7.6. Node numbering for the barrel vault*

*Check for deflection*

The maximum deflection of 84.57 mm was found to occur at node 181 under case 1 loading. This is at the ultimate limit state. The deflection at the serviceability limit state may be computed as 84.7/1.50 = 56.37 mm. The Indian Code, IS 800: 1984, restricts deflections to span/325, which is more stringent than the limit prescribed in BS 5950: Part 1: 1985. In this instance, the permissible deflection is 22 900/325 = 70.46 mm. The actual maximum deflection of 56.37 mm is less than what is permissible.

## *7.6.3. Typical design and detailing of a joint*

We consider here the design and detailing of a joint between the top chords of the space frame and the rectangular hollow section edge member. For the purposes of illustration, node 58, where top chord members 55, 721 and 739 meet, is selected. Members 55, 721 and 739 connect node 58 to nodes 59, 40 and 78, respectively. A plate 220 mm wide and 10 mm thick will be welded to the top of edge member consisting of a 132 mm × 132 mm × 5.4 mm rectangular hollow section. Node 58 is conceptually located at the centre of the rectangular hollow section.

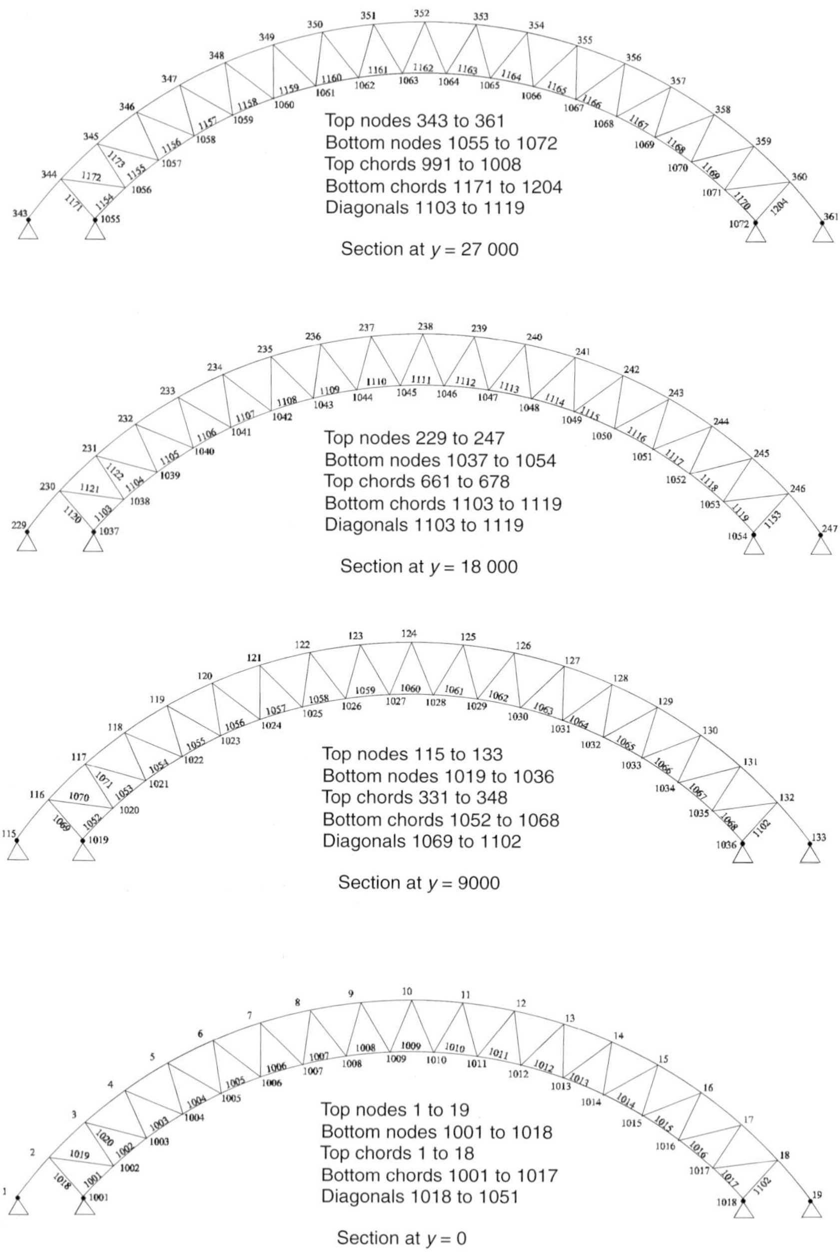


*Fig. 7.7. Node and member numbering for the stiffening arches*

*Table 7.1. The global and local co-ordinates for the nodes*

| Node No. | Global co-ordinates | | | Local co-ordinates | | |
|---|---|---|---|---|---|---|
| | $x$ | $y$ | $z$ | $X$ | $Y$ | $Z$ |
| 58 | −11 450.0000 | 4500 | 8 525.2174 | 0 | 0 | 0 |
| 59 | −10 508.7085 | 4500 | 9 661.7222 | 941.2915 | 0 | 1136.5048 |
| 40 | −10 508.7085 | 3000 | 9 661.7222 | 941.2915 | −1500 | 1136.5048 |
| 78 | −10 508.7085 | 6000 | 9 661.7222 | 941.2915 | 1500 | 1136.5048 |

*Table 7.2. The direction cosines and lengths for the members*

| Member No. | Length | cos $\alpha$ | cos $\beta$ | cos $\gamma$ |
|---|---|---|---|---|
| 55 | 1475.6940 | 0.6379 | 0 | 0.7701 |
| 739 | 2104.2036 | 0.4473 | 0.7128 | 0.5401 |
| 721 | 2104.2036 | 0.4473 | −0.7128 | 0.5401 |

*Step 1*

Choosing node 58 as the origin, the global and local co-ordinates of nodes 59, 40 and 78 are tabulated in Table 7.1.

*Step 2*

The lengths of members 55, 739 and 721 and their direction cosines, cos $\alpha$, cos $\beta$ and cos $\gamma$, with respect to the $x$, $y$ and $z$ axes, respectively, are given in Table 7.2.

*Step 3*

In this step we determine the major and minor axes. The sizes of the tubular members 55, 739 and 721 are 48.30 mm × 3.25 mm, 114.30 mm × 4.50 mm and 114.30 mm × 4.50 mm, respectively. The major and minor axes of their elliptical traces on the plate to which they are welded are computed. For example, member 55 exits the top plate with a direction cosine of 0.6379 to the $x$ axis corresponding to an angle of 50.37°.

Referring to Fig. 7.8, the major axis of the elliptical trace of the tube on the plate is given by 48.30/sin 50.37 = 62.71 mm. The minor axis is 48.30 mm.

Member 721 exits the plate at $Z = 76$ mm. The $X$ and $Y$ co-ordinates of the exit point may be worked out as follows:

$$\begin{aligned}\text{Length up to the exit point} &= 76/\cos\gamma \\ &= 76/0.5401 \\ &= 140.71 \text{ mm}\end{aligned}$$

Hence,

$$\begin{aligned}X &= 140.71 \times \cos\alpha \\ &= 140.71 \times 0.4473 \\ &= 62.94 \text{ mm}\end{aligned}$$

$$\begin{aligned} Y &= 140.71 \times \cos \beta \\ &= 140.71 \times -0.7128 \\ &= -100.30 \text{ mm} \end{aligned}$$

The exit point on the plate is at C (Fig. 7.9). The length AC is given by

$$\sqrt{62.94^2 + 100.30^2} = 118.41 \text{ mm}$$

The line OC makes an angle $\theta$ to the plate:

$$\theta = \cos^{-1}\left(\frac{118.41}{140.71}\right) = 32.70°$$

Hence the major axis of the trace of member 721 of diameter 114.30 mm is

$$\frac{114.30}{\sin 32.70} = 211.57 \text{ mm}$$

The minor axis is 114.30 mm.

The positions of the exit points of the three tubes and their elliptical traces are shown in Fig. 7.9.

### *Design of welds*

By way of illustration, the welded connection between member 55 and the plate will be designed.

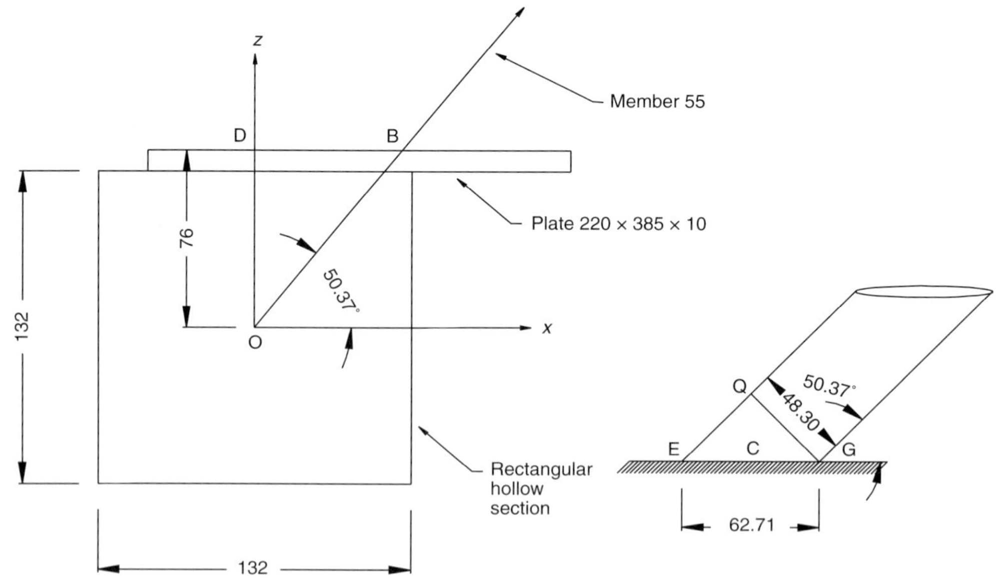


*Fig. 7.8. Details of the joint*

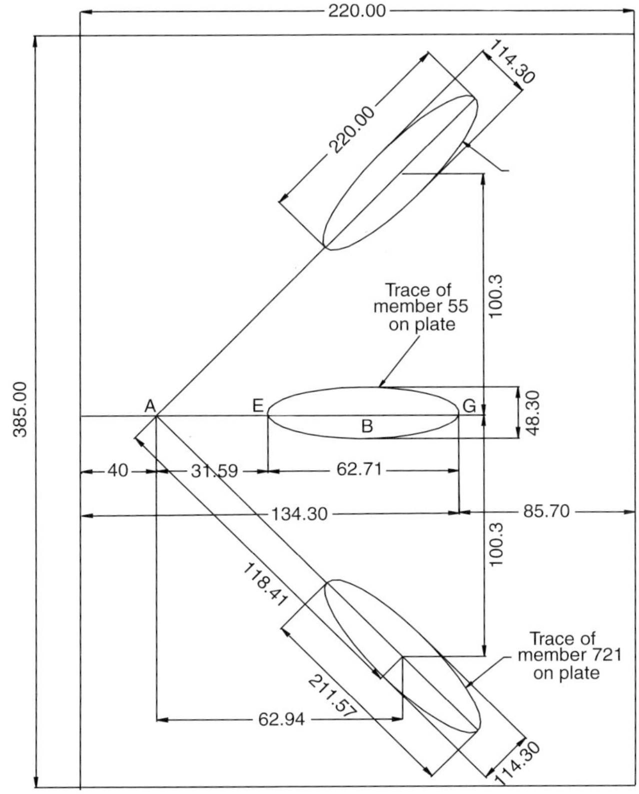


*Fig. 7.9. The traces for the tubes attached to the plate*

The length of the weld is equal to the perimeter of the ellipse, with a semi-major axis $a = 31.36$ mm and a semi-minor axis $b = 24.15$ mm, which may be approximated as $\pi(a + b) = 174.35$.

Assuming a 4 mm weld with a yield strength of 350 MPa, the strength of the weld is

$$\frac{174.35 \times 4 \times 350}{1000} = 244 \text{ kN}$$

The force in the member at the ultimate limit state is 45.93 kN tension. Hence, the weld is safe.

## 7.6.4. *Calculationof drilling angles*

The calculation of the drilling angles is here illustrated by means of a step by step procedure for node 2.

*Step 1*

First we must identify the nodes to which node 2 is connected. These are 1, 3, 1001, 1002, 20 and 21. The global co-ordinates for these nodes are tabulated in Table 7.3.

*Step 2*

Choosing node 2 as the origin, we first tabulate the local co-ordinates $X_T$, $Y_T$ and $Z_T$ of the nodes involved (Table 7.4). If the global co-ordinates $x$, $y$, $z$ of the reference node are denoted by $A$, $B$ and $C$. The local co-ordinates of the other nodes to which node 2 is connected are found as:

$$X_T = x - A, \qquad Y_T = y - B, \qquad Z_T = z - C \tag{7.1}$$

*Step 3*

Choosing the origin to be at node 2, the line joining nodes 2 and 3 to be the axis $X_N$, and a line passing at a right angle through the origin O to be the axis $Z_N$, we can write the new co-ordinates $X_N$, $Y_N$ and $Z_N$ using the following transformation (Fig. 7.10):

$$X_N = X_T \cos \Delta + Z_T \sin \Delta \tag{7.2}$$

$$Z_N = -X_T \sin \Delta + Z_T \cos \Delta \tag{7.3}$$

where, $\Delta$ is the angle made by the line joining nodes 2 and 3 to the horizontal. The angle $\Delta$ is found from Fig. 7.11 as 44.44°.

To facilitate the calculation of $Z_N$ and $X_N$, the angles $\Delta$ made by lines joining each node $n$ on the extrados to the next node $n + 1$ are tabulated (Table 7.5). Noting that

$$\cos \Delta = \cos 44.44 = 0.713984048$$

and

*Table 7.3. The global co-ordinates for the nodes*

| Node No. | $x$ | $y$ | $z$ |
|---|---|---|---|
| 2 | −10 508.7085 | 0 | 9 661.7222 |
| 1 | −11 450.0000 | 0 | 8 525.2174 |
| 3 | −9 455.1174 | 0 | 10 694.9790 |
| 1001 | −9 514.5598 | 0 | 8 525.2174 |
| 1002 | −8 944.9982 | 0 | 9 121.0299 |
| 20 | −11 450.0000 | 1500 | 8 525.2174 |
| 21 | −10 508.7085 | 1500 | 9 661.7222 |

*Table 7.4. The local co-ordinates for the nodes*

| Node No. | $X_T$ | $Y_T$ | $Z_T$ |
|---|---|---|---|
| 2 | 0 | 0 | 0 |
| 1 | −941.2915 | 0 | −1136.5048 |
| 3 | 1053.5906 | 0 | 1033.2568 |
| 1001 | 994.1487 | 0 | −1136.5048 |
| 1002 | 1563.7103 | 0 | −540.6923 |
| 20 | −941.2915 | 1500 | −1136.5048 |
| 21 | 0 | 1500 | 0 |

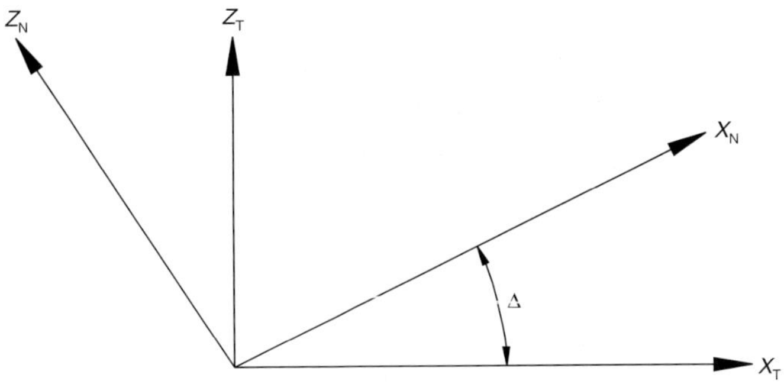


*Fig. 7.10. Rotation of the co-ordinates*

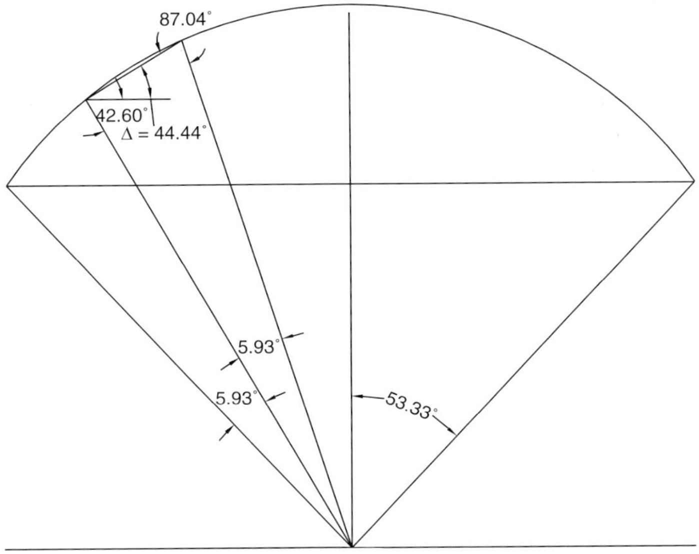


*Fig. 7.11. The angle Δ*

$$\sin \Delta = \sin 44.44 = 0.700161966$$

the values of $X_N$ and $Z_N$ are found and tabulated (Table 7.6). In Table 7.6 $L_N$ is the length of the member joining node 2 to the node listed.

*Step 4*

With the help of Table 7.6, the position of the holes to be drilled in node 2 can be easily located. The origin is at the centre of the node to be drilled for the $X_N$, $Z_N$ Cartesian co-ordinates. Holes drilled at node 2 are for connecting tubular members the remote ends of which are at nodes 1, 3, 1001, 1002, 20 and 21. The holes will be designated by 1, 3, 1001, 1002, 20 and 21, respectively. Members 2–1, 2–3, 2–1001 and 2–1002 lie in the $X_N$, $Z_N$ plane and hence the values of $X_N$ and $Z_N$ are adequate to locate these holes. The values for $\phi$ for locating the holes are computed as follows.

For hole 1:

$$\phi = \tan^{-1}\left(\frac{-152}{-1467.801}\right) = 5.93°$$

$\phi$ is in the third quadrant and is equal to 95.93°, measured anticlockwise from the positive $X_N$ axis.

For hole 1001:

$$\phi = \tan^{-1}\left(\frac{-1507.514}{-85.89404}\right) = 86.74°$$

Because both $Z_N$ and $X_N$ are negative, $\phi$ lies in the third quadrant and is equal to 176.74°, measured anticlockwise from the positive $X_N$ axis.

For hole 1002:

$$\phi = \tan^{-1}\left(\frac{-1480.879}{737.9284}\right) = 63.51°$$

*Table 7.5. The values of Δ for the node combinations*

| Node $n$ | Node $n+1$ | Angle to the horizontal of the line joining $n$ and $n+1$ (°) | $\Delta$ (rad) |
|---|---|---|---|
| 1 | 2 | 50.37 | 0.8791 |
| 2 | 3 | 44.44 | 0.7756 |
| 3 | 4 | 38.51 | 0.6721 |
| 4 | 5 | 32.58 | 0.5686 |
| 5 | 6 | 26.65 | 0.4651 |
| 6 | 7 | 20.72 | 0.3616 |
| 7 | 8 | 14.79 | 0.2581 |
| 8 | 9 | 8.86 | 0.1546 |
| 9 | 10 | 2.93 | 0.0511 |

*Table 7.6. The Cartesian co-ordinates after rotation*

| Node No. | $X_N$ | $Y_N$ | $Z_N$ | $L_N$ |
|---|---|---|---|---|
| 2 | – | – | – | – |
| 1 | –1467.805 | 0 | –152.3898 | 1475.694 |
| 3 | 1475.694 | 0 | 0 | 1475.694 |
| 1001 | –85.89404 | 0 | –1507.514 | 1509.959 |
| 1002 | 737.9284 | 0 | –1480.879 | 1654.551 |
| 20 | –1467.801 | 1500 | –152.3898 | 2104.203 |
| 21 | 0 | 1500 | 0 | 1500 |

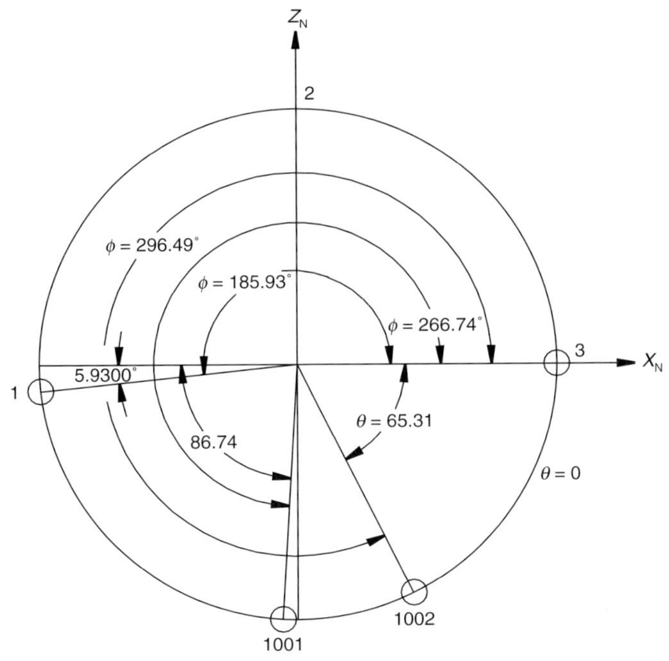


*Fig. 7.12. The values of θ and φ for the drill holes for hole 1002*

Because $Z_N$ is positive and $X_N$ is negative, $\phi$ lies in the fourth quadrant and is equal to 63.51°, measured clockwise from the positive $X_N$ axis. The location of the drill holes is shown in Fig. 7.12.

Hole 20 is not in the $X_N$, $Z_N$ plane, and hence it is located using the spherical co-ordinates $\phi$ and $\theta$, which are measured in a counter-clockwise direction from the axes $Z_N$ and $X_N$, respectively. We have the following relationships:

$$\phi = \cos^{-1}\left(\frac{Z_N}{L_N}\right) \tag{7.4}$$

and

$$\tan\theta = \left(\frac{Y_N}{X_N}\right) \tag{7.5}$$

From Table 7.4, for node 20,

$$\left(\frac{Z_N}{L_N}\right) = \left(\frac{-152.3898}{2104.203}\right)$$

Hence $\phi = 94.15°$ measured anti-clockwise from the positive direction of the $Z_N$ axis.

$$\left(\frac{Y_N}{X_N}\right) = \left(\frac{1500}{-1467.801}\right)$$

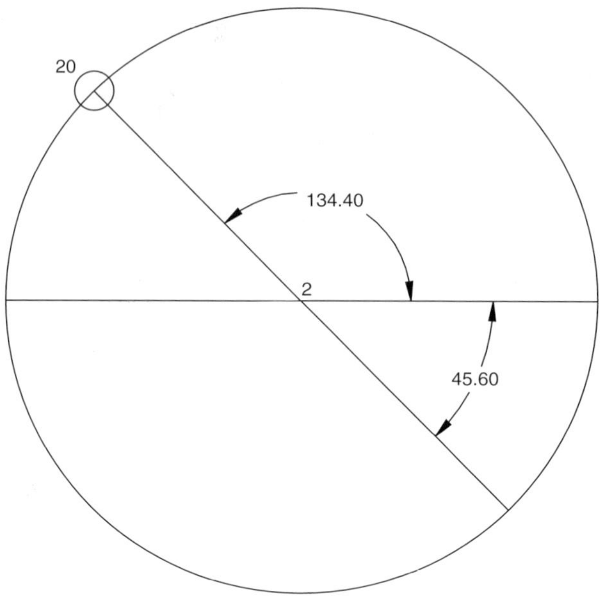


*Fig. 7.13. The value of θ for the drill hole 20*

*Table 7.7. The values of θ and ϕ for the drill holes at different nodes*

| Node No. | $\theta$ (°) | $\phi$ |
|---|---|---|
| 3 | 0 | 0 |
| 1 | 0 | 185.93 |
| 1001 | 0 | 266.74 |
| 1002 | 0 | 296.49 |
| 20 | 134.4 | 94.15 |
| 21 | 90 | 0 |

Hence $\theta = -45.6°$ or 134.4°. It is quite clear from the physical location of node 20 with reference to node 2 that the appropriate value is 134.4° (Fig. 7.13). The $\phi$ and $\theta$ values for the drill holes are summarized in Table 7.7.

### 7.6.5. Credits

The owner of Thirumailai Station is the Southern Railway and the contractor for the job was Larsen and Toubro Ltd.

## 7.7. References

7.1. MAKOWSKI, Z. S. (ed.). History of development of various types of braced vaults and review of recent developments all over the world. In: *Analysis, Design and Construction of Barrel Vaults*, Applied Science Publishers, London, 1985, chap. 1, pp. 1–35.

7.2. RAMASWAMY, G. S. *Design & Construction of Concrete Shell Roofs*, McGraw–Hill, New York, 1968, chap. 4, pp. 45–49.
7.3. MAKOWSKI, Z. S. (ed.). History of development of various types of barrel vaults and review of recent developments all over the world. In: *Analysis, Design and Construction of Braced Barrel Vaults*, Applied Science, London, 1985, chap. 1, pp. 7–13.
7.4. PARKE, G. A. R. Comparison of the structural behaviour of various types of braced barrel vaults. In: Z. S. Makowski (ed.), *Analysis, Design and Construction of Braced Barrel Vaults*, Allied Science, London, 1985, chap. 6, pp. 91–120.
7.5. WALKER, H. B. Basic principles in the design and construction of braced barrel vault roofs. In: Z. S. Makowski (ed.), *Analysis, Design and Construction of Braced Barrel Vaults*, Applied Science, London, 1985, chap. 11, pp. 197–207.
7.6. DU CHATEÂU, S. Architectural aspects of braced barrel vaults. In: Z. S. Makowski (ed.), *Analysis, Design and Construction of Braced Barrel Vaults*, Allied Science, London, 1985, chap. 14, pp. 223–227.

# 8. Braced domes

## 8.1. Introduction

The dome has always appealed to engineers and architects as an attractive and structurally efficient solution for covering large unobstructed areas by virtue of its ability to enclose a given volume with the least surface area. When used to roof stadia, it provides unobstructed sight lines for a large number of spectators. Makowski has traced the evolution of braced domes and classified them on the basis of the bracing systems used. He has reviewed current developments and listed some of the remarkable braced domes built in recent years [8.1].

## 8.2. The relative merits of different bracing schemes

Richter [8.2] has presented the results of a study carried out to evaluate the relative merits of Schwedler, lattice, lamella and geodesic domes. Their relative strength and stiffness characteristics are given in Fig. 8.1. He concluded that the geodesic dome has an edge over the others, in terms of both strength and stiffness.

## 8.3. Platonic polyhedra

There are only five polyhedra that have equal sides: the tetrahedron, the hexahedron or the cube, the octahedron, the dodecahedron and the icosahedron. They belong to two families. The first family consists of polyhedra the faces of which are equilateral triangles (tetrahedron, octahedron and icosahedron), and the second comprises polyhedra with only three edges meeting at a vertex (tetrahedron and hexahedron). The tetrahedron belongs to both families. The number of vertices $V$, edges $E$ and faces $F$ are related by the equation $V - E + F = 2$, which is known as *Euler's formula*.

## 8.4. The geodesic dome

The geodesic dome owes its origin and development to the pioneering work of Buckminster Fuller [8.3]. Being the largest of the regular platonic polyhedra, approaching the sphere more closely than the others, Fuller chose the spherical icosahedron as the logical point of departure in his search for the optimum bracing scheme for enclosing space. The icosahedron has 30 edges, 20 faces (all equilateral

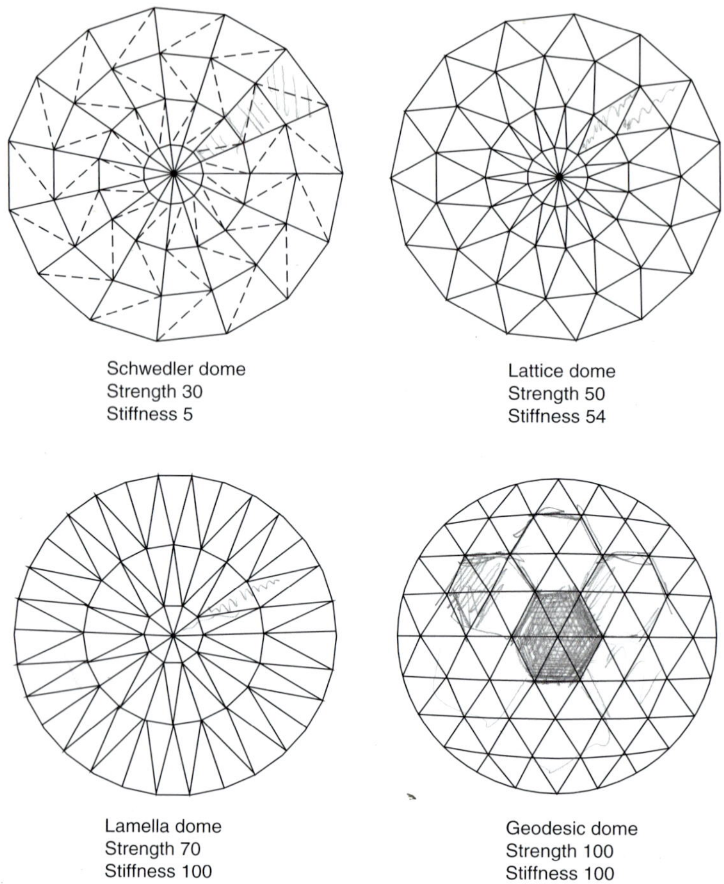


*Fig. 8.1. The structural efficiencies of different bracing systems. (Reproduced with permission from Richter [8.2])*

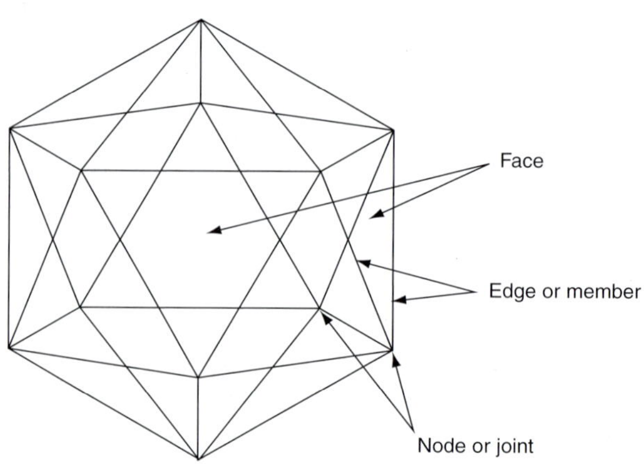


*Fig. 8.2. An icosahedron*

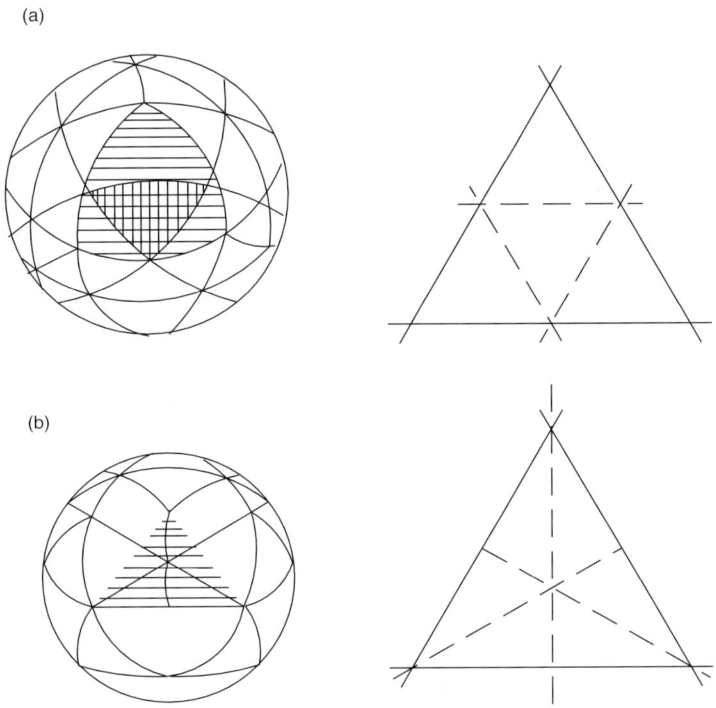


*Fig. 8.3. The two ways of orienting the three-way grid: (a) class I, alternate breakdown; (b) class II, triacon breakdown*

triangles) and 12 vertices (Fig. 8.2). The length of each edge is

$$\frac{1}{5}R\sqrt{10(5-\sqrt{5})} = 1.0515R$$

where $R$ is the radius of the circumscribing sphere. In domes of large radius, the length of the edges becomes very long and buckling will govern their design. The members will, therefore, have to be very heavy. Obviously, the edges of the icosahedron have to be subdivided to provide additional vertices on the sphere in order to keep the lengths and sizes of the members within manageable limits. After this subdivision, the edges of the resulting polyhedron will no longer be equal. The aim should, therefore, be to effect the subdivision in such a manner that the number of different lengths is kept to a minimum. Fuller's ingenious scheme of subdivision led to the development of the geodesic dome.

## 8.5. Subdivision schemes

The triangular polyhedral faces must first be subdivided by a three-way grid. All the vertices of this grid have to be pushed until they lie on the sphere. There are two ways of orienting the grid (Fig. 8.3).

In the *class I (alternate) scheme* the grid lines run parallel to the sides of the polyhedral triangular face to create subtriangles. In the *class II (triacon) scheme* the orientation of the division lines is perpendicular to the triangular edges of the polyhedral triangular faces comprising the icosahedron. In both cases, the number of parts into which the edge is subdivided is known as the *frequency*. The relative merits of the two schemes may be summed up as follows:

- class I subdivision permits both odd and even frequencies
- class I subdivision results in uninterrupted equators at all even frequencies
- class I subdivision keeps the variations in the lengths of the components within narrow bounds
- class I subdivision involves a larger inventory of different lengths
- although class II subdivision results in a smaller inventory of different lengths, the differences between the lengths is larger, and hence the stress distribution in class II domes is less uniform.

The next step involves the breakdown of the subdivided polyhedral faces. Two methods are commonly employed.

*Method 1* involves drawing the icosahedron face on a flat sheet of paper, dividing the triangular edges equally and connecting them to form a triangular grid (Fig. 8.4). The division points are projected to lie on the sphere. It is to be noted that while AC = CD = DB, AF, FG and GB are not equal, the edge chords being shorter than the middle chord. Being points on the sphere, OA, OF, OG and OB are equal in length (Fig. 8.4).

M*ethod 2* involves projecting the division points onto the sphere to make the angles AOF, FOG and GOB equal (Fig. 8.4). Consequently, the chords AF, FG and GB are equal.

In the examples that follow we used the class I, method 1 breakdown, because of the many advantages that it offers.

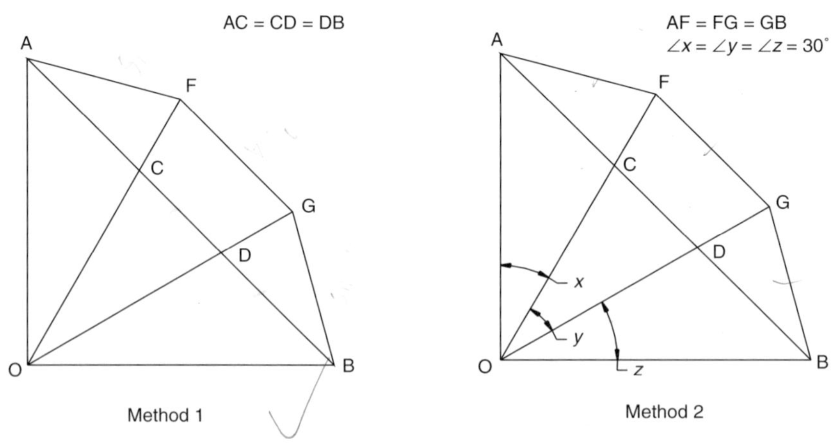


*Fig. 8.4. The methods for breaking down the subdivided polyhedral faces. (Reproduced with permission from Kenner [8.4])*

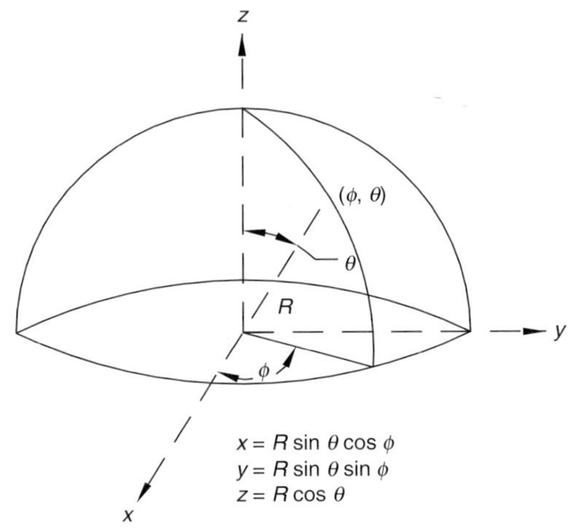


*Fig. 8.5. The spherical co-ordinates*

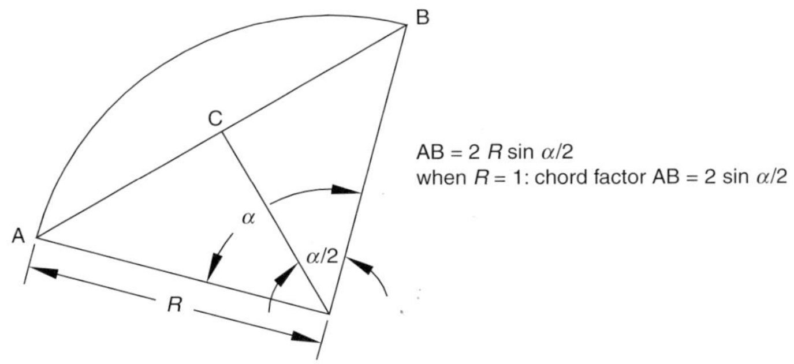


*Fig. 8.6. Definition of the chord factor*

## 8.6. Spherical co-ordinates

Calculations relating to geodesic domes are best handled by using spherical co-ordinates (Fig. 8.5). It is easy to show that

$$x = R \sin \theta \cos \phi \tag{8.1a}$$

$$y = R \sin \theta \sin \phi \tag{8.1b}$$

$$z = R \cos \theta \tag{8.1c}$$

## 8.7. Chord factors

Referring to Fig. 8.6, the distance AB is equivalent to $2R \sin(\alpha/2)$ and the chord factor is $2 \sin(\alpha/2)$. If we know the spherical co-ordinates $\phi_1$, $\theta_1$, $R_1$ and $\phi_2$, $\theta_2$, $R_2$ of two points on two domes of radii $R_1$ and $R_2$, the distance $d$ between them is given by

$$d = \sqrt{R_1^2 + R_2^2 - 2R_1 2R_2[\cos\theta_1 \cos\theta_2 + \cos(\phi_1 - \phi_2)\sin\theta_1 \sin\theta_2]} \tag{8.2}$$

If the points are on the same sphere, then $R_1 = R_2$; and if $R_1 = R_2 = 1$, $d$ becomes the chord factor, and we may write

$$d = \sqrt{2 - 2[\cos\theta_1 \cos\theta_2 + \cos(\phi_1 - \phi_2)\sin\theta_1 \sin\theta_2]} \tag{8.3}$$

A computer program for calculating chord factors, CHORD, is included on the accompanying CD-ROM. To find the distance between two points on the sphere, the chord factor is multiplied by its radius.

## 8.8. Derivation of co-ordinates

The vertices on the triangular face of an icosahedron resulting from class I subdivision are most conveniently represented by a set of three integers $L$, $M$ and $N$. Referring to Fig. 8.7, at the point P, $L = 3$, $M = 1$ and $N = 2$. Hence the co-ordinates of P are 3, 1, 2. It may be noted that

$$L + M + N = f \tag{8.4}$$

where $f$ is the frequency.

Once the values of $L$, $M$ and $N$ for a point are available, we can derive the spherical and Cartesian co-ordinates as follows:

$$L_1 = L \sin 72 \tag{8.5a}$$

$$M_1 = M + L \cos 72 \tag{8.5b}$$

$$N_1 = \frac{f}{2} + \frac{N}{2\cos 36} \tag{8.5c}$$

$$\phi_1 = \tan^{-1}\left(\frac{L_1}{M_1}\right) \tag{8.6a}$$

$$\theta_1 = \tan^{-1}\left(\frac{\sqrt{M_1^2 + L_1^2}}{N_1}\right) \tag{8.6b}$$

$$x = R \sin\theta_1 \cos\phi_1 \tag{8.7a}$$

$$y = R \sin\theta_1 \sin\phi_1 \tag{8.7b}$$

$$z = R \cos\theta_1 \tag{8.7c}$$

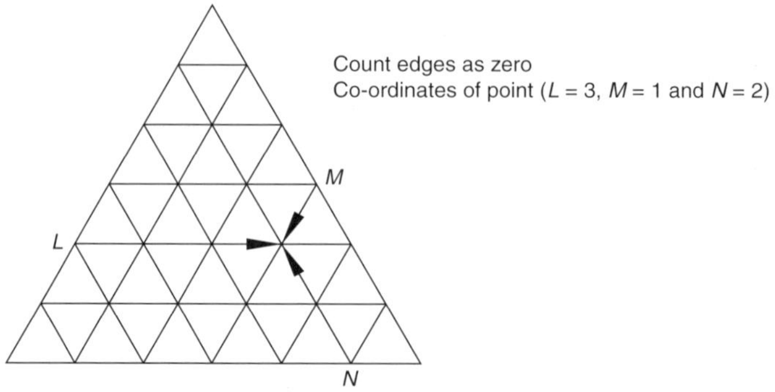


*Fig. 8.7. The method of three integers* L, M *and* N. *(Reproduced with permission from Kenner [8.4])*

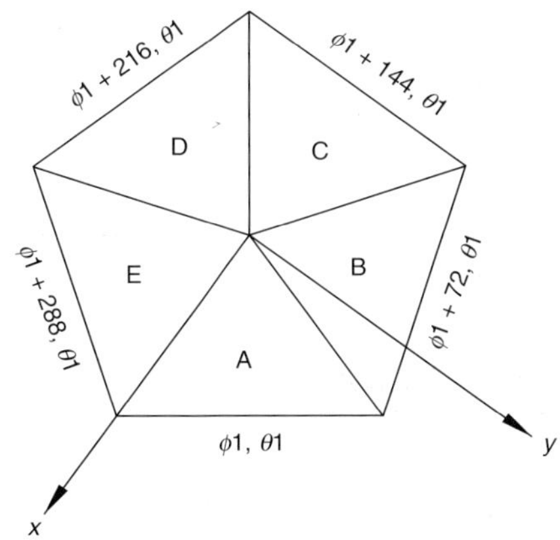


*Fig. 8.8. The spherical and Cartesian co-ordinates*

The above derivations are based on the work of Kenner [8.4].

## 8.9. Generating the dome topology

The top five faces of the icosahedron surrounding the vertex form what is known as the *pentacap*, the rise/span ratio of which can be shown to be 0.3090 [8.4]. The topology of most shallow domes occurring in practice can be found within the pentacap and the illustrative examples that follow belong to this class. For the method for generating the topology of the other geodesic domes the reader is referred to Kenner [8.4]. We have shown how the spherical and Cartesian co-ordinates of one of the faces, labelled A in Fig. 8.8, can be generated. For triangles B, C, D and E the values of $\phi$ are incremented as follows:

- triangle B, $\phi + 72°$
- triangle C, $\phi + 144°$
- triangle D, $\phi + 216°$
- triangle E, $\phi + 288°$.

The values of $\theta$ remain unchanged. Note that $\theta = \tan^{-1} 2 = 63.43°$ at the base of the pentacap.

## 8.10. Identifiers

The identifiers $U_1$ and $U_2$ are introduced to facilitate the numbering of nodes. They are related to $L$, $M$ and $N$ as follows:

$$L = U_2 \tag{8.8a}$$

$$M = U_1 - U_2 \tag{8.8b}$$

$$N = f - U_1 \tag{8.8c}$$

The concept of identifiers may be easily understood by referring to Fig. 8.9, which shows an icosahedral face subdivided into a frequency of 4. A vertex labelled 3, 2 will mean that for that vertex $U_1 = 3$ and $U_2 = 2$. Let us consider the vertex 3, 2. Its $L$, $M$ and $N$ co-ordinates are $L = 2$, $M = 1$ and $N = 1$. It is now easy to see that the relationships (8.8a) to (8.8c) are satisfied.

## 8.11. Computer program

A computer program, GEODOME, for generating the co-ordinates of all shallow domes with a rise/span ratio of less than 0.3090 is included on the accompanying CD-ROM. The value of $K$ is successively input as 0, 1, 2, 3 and 4 to generate co-ordinates of vertices on the faces A, B, C, D and E (Fig. 8.8), respectively.

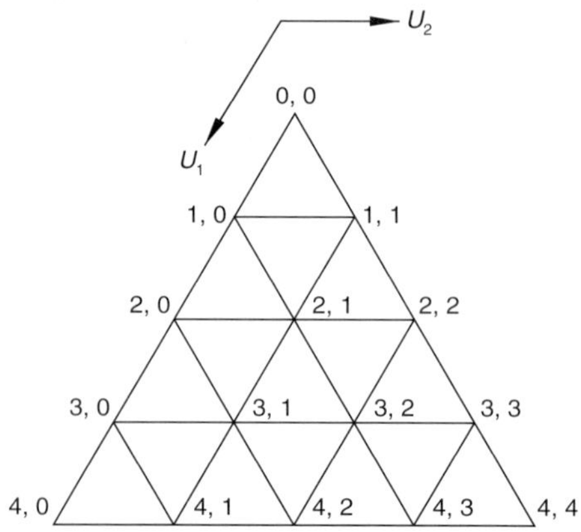


*Fig. 8.9. The identifiers for an icosahedral face subdivided into a frequency of 4*

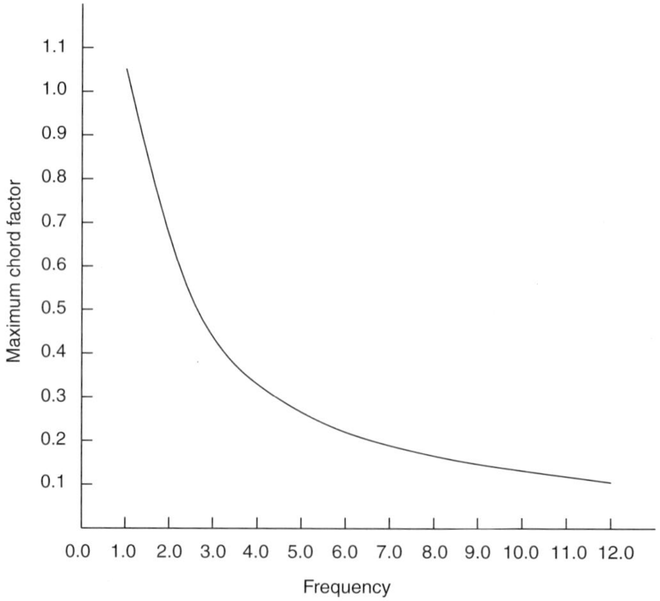


*Fig. 8.10. The variation of the chord factor with frequency*

*Table 8.1. The maximum chord factors for given frequencies*

| Frequency | Maximum chord factor |
|---|---|
| 1.0 | 1.0515 |
| 2.0 | 0.6180 |
| 3.0 | 0.4124 |
| 4.0 | 0.3249 |
| 5.0 | 0.2616 |
| 6.0 | 0.2166 |
| 7.0 | 0.1879 |
| 8.0 | 0.1646 |
| 9.0 | 0.1458 |
| 10.0 | 0.1319 |
| 12.0 | 0.1054 |

## 8.12. Selection of the frequency

The basis for the selection of the appropriate frequency is the maximum acceptable length of members. The maximum length of members in the dome is the product of the radius of the dome and the maximum chord factor. It can be seen from Fig. 8.10 and Table 8.1 that the maximum chord factors vary inversely with respect to the frequency. Hence, high frequencies have to be selected for domes of large diameter in order to keep the maximum size of the members within manageable limits so that buckling considerations do not make them unduly heavy. However, higher frequencies will lead to an increase in the number of members and they will also result in an increase in the variety of member lengths. In general, one should select the lowest practicable frequency in order to minimize the component inventory. In single-layer domes, which are vulnerable to snap-through buckling, it is a good rule to avoid frequencies higher than 12.

## 8.13. Truncation

When the span and rise of a dome are known, the radius can be computed. For example, if the span is 12 m and the rise is 3 m, the radius works out as 7.5 m (Fig. 8.11). If the maximum acceptable length for a member is restricted to 2.5 m, the appropriate frequency may be selected from Table 8.1. If the frequency chosen is 4, the maximum chord factor is 0.3249 and the maximum length of the member is $0.3249 \times 7.5 = 2.4367$ m. As this value is less than the prescribed limit of 2.5 m the frequency of 4 is appropriate.

Truncation is a technique used to restrict the dome rise as closely as possible to the prescribed 3 m rise and to achieve a flat circular plan at its bottom so that it can be supported on a ring beam. The concepts of frequency selection, truncation and dome topology generation are best illustrated by means of an example.

*Example 8.1*

A geodesic dome has a span of 12 m and a rise of 3 m. The maximum length of the members in the dome is restricted to 2.5 m and the dome is to be within the pentacap. Our task here is to generate the dome topology.

*Solution.* It has already been shown that the appropriate frequency for this dome is 4. Using the program GEODOME and setting $R = 7500$ and $K = 0$, the co-ordinates of the points on the sphere corresponding to the intersection points of the tri-linear grid on the triangular icosahedral face A are generated (Table 8.2).

Inspection of Table 8.2 shows that the points $U_1 = 4$ and $U_2 = 0$, $U_1 = 4$ and $U_2 = 1$, $U_1 = 4$ and $U_2 = 2$, $U_1 = 4$ and $U_2 = 3$, and $U_1 = 4$ and $U_2 = 4$ correspond to a rise of the dome in excess of the prescribed 3 m. Hence the dome needs to be terminated at $U_1 < 4$ or it needs to be truncated.

Before we decide the truncation level, the value of $\theta$ corresponding to the given rise of 3 m is computed from Fig. 8.11 as follows:

$$\theta = \sin^{-1}\left(\frac{6}{7.5}\right) = 53.13°$$

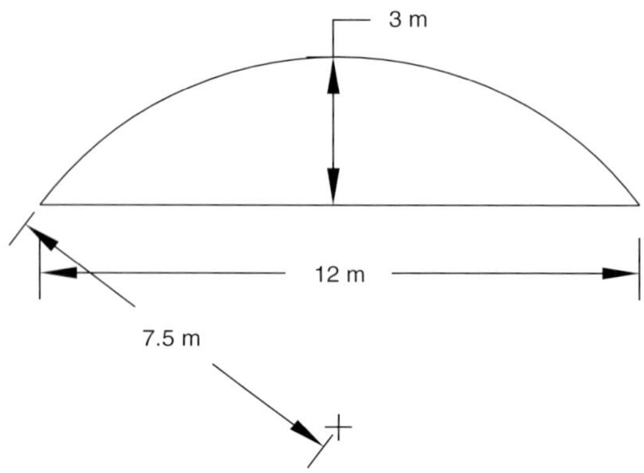


*Fig. 8.11. Calculation of the radius*

*Table 8.2. The co-ordinates of the points on the sphere*

| Serial No. | $U_1$ | $U_2$ | $\theta$ (rad) | $\phi$ (rad) | $x$ (mm) | $y$ (mm) | $z$ (mm) |
|---|---|---|---|---|---|---|---|
| 1 | 1 | 0 | 0.2538657 | 0 | 1883.608 | 0 | 7259.616 |
| 2 | 1 | 1 | 0.2538657 | 1.256637 | 582.0667 | 1791.417 | 7259.616 |
| 3 | 2 | 0 | 0.5535744 | 0 | 3942.983 | 0 | 6379.881 |
| 4 | 2 | 1 | 0.4636476 | 0.6283185 | 2713.525 | 1971.492 | 6708.204 |
| 5 | 2 | 2 | 0.5535744 | 1.256637 | 1218.449 | 3750 | 6379.881 |
| 6 | 3 | 0 | 0.853283 | 0 | 5650.823 | 0 | 4931.349 |
| 7 | 3 | 1 | 0.7617826 | 0.3907125 | 4786.475 | 1971.492 | 5427.051 |
| 8 | 3 | 2 | 0.7617826 | 0.8659245 | 3354.102 | 3942.984 | 5427.051 |
| 9 | 3 | 3 | 0.853283 | 1.256637 | 1746.2 | 5374.252 | 4931.349 |
| 10 | 4 | 0 | 1.107149 | 0 | 6708.204 | 0 | 3354.102 |
| 11 | 4 | 1 | 1.044545 | 0.27987 | 6232.89 | 1791.417 | 3767.215 |
| 12 | 4 | 2 | 1.017222 | 0.6283185 | 5161.432 | 3750 | 3942.984 |
| 13 | 4 | 3 | 1.044545 | 0.976767 | 3629.808 | 5374.252 | 3767.215 |
| 14 | 4 | 4 | 1.107149 | 1.256637 | 2072.949 | 6379.881 | 3354.102 |

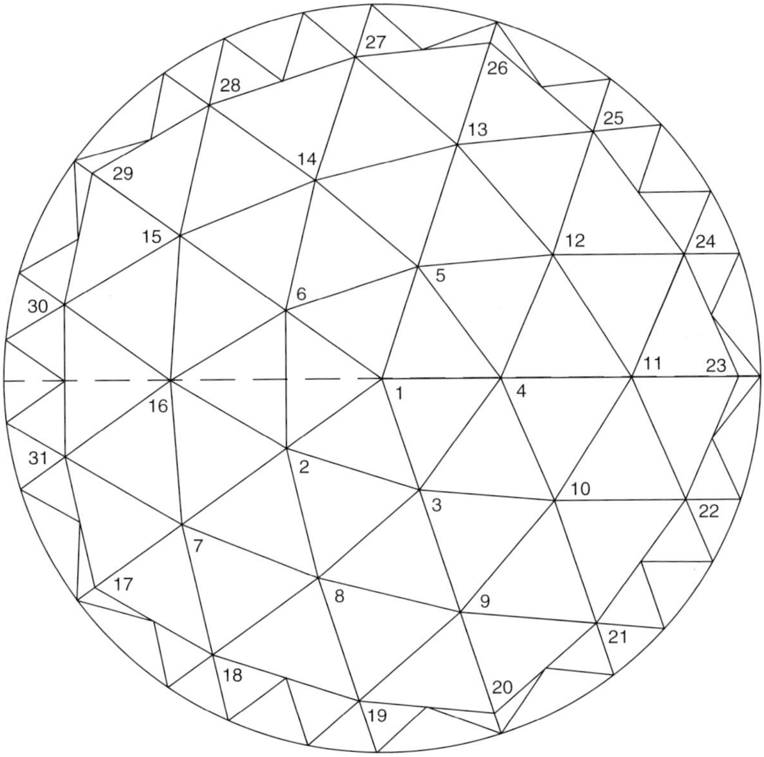


*Fig. 8.12. Plan of the dome*

Expressed in radians, $\theta = 0.9273$. The radius of the latitudinal circle corresponding to the location of the ring beam is computed as

$$R \sin \theta = 7500 \sin 53.13 = 6000 \text{ mm}$$

The rise of the dome is then

$$R(1 - \cos \theta) = 7500(1 - \cos 53.13) = 3000 \text{ mm}$$

Because the computed value of $\theta$ (53.13°) is less than the prescribed value of 63.43°, the dome will be entirely within the pentacap. The truncation level is chosen as $l = 3$. The dome is stopped at this level and the bottom-most nodes corresponding to $U_1 = 3$ and $U_2 = 0$, $U_1 = 3$ and $U_2 = 1$, and $U_1 = 3$ and $U_2 = 3$ will lie above the ring beam and will have to be connected to it by triangulating the gap (Fig. 8.12). By setting $K$ successively as 1, 2, 3 and 4, the co-ordinates for the entire dome with a frequency of 4 and a truncation level of 3 can be generated (Table 8.3). A perspective view of the dome is given in Fig. 8.13.

*The number of joints in the dome*

If $l < f$, the number of joints $N_J$ in the dome may be found from the rule

$$N_J = \sum_{p=0}^{p=l} [5p+5] - (5l+4) \tag{8.9}$$

For the dome in the present example $l = 3$ and thus $N_J = 31$.

*The number of members in the dome*

$$N_M = \sum_{p=1}^{p=l} [15p-5] \tag{8.10}$$

For the dome in the present example $l = 3$ and thus $N_M = 75$.

*Table 8.3. The co-ordinates for the dome (frequency 4, truncation level 3)*

| Node No. | $x$ (mm) | $y$ (mm) | $z$ (mm) |
|---|---|---|---|
| 1 | 0 | 0 | 0 |
| 2 | 1883.608 | 0 | 7259.616 |
| 3 | 582.0667 | 1791.417 | 7259.616 |
| 4 | −1523.871 | 1107.157 | 7259.616 |
| 5 | −1523.871 | −1107.157 | 7259.616 |
| 6 | 582.0667 | −1791.417 | 7259.616 |
| 7 | 3942.983 | 0 | 6379.881 |
| 8 | 2713.525 | 1971.492 | 6708.204 |
| 9 | 1218.449 | 3750 | 6379.881 |
| 10 | −1036.475 | 3189.94 | 6708.204 |
| 11 | −3189.94 | 2317.627 | 6379.881 |
| 12 | −3354.102 | 0 | 6708.204 |
| 13 | −3189.94 | −2317.628 | 6379.881 |
| 14 | −1036.476 | −3189.94 | 6708.204 |
| 15 | 1218.449 | −3750 | 6379.881 |
| 16 | 2713.525 | −1971.492 | 6708.204 |
| 17 | 5650.823 | 0 | 4931.349 |
| 18 | 4786.475 | 1971.492 | 5427.051 |
| 19 | 3354.102 | 3942.984 | 5427.051 |
| 20 | 1746.200 | 5374.252 | 4931.349 |
| 21 | −395.8982 | 5161.432 | 5427.051 |
| 22 | −2713.526 | 4408.39 | 5427.051 |
| 23 | −4571.612 | 3321.47 | 4931.349 |
| 24 | −5031.153 | 1218.448 | 5427.051 |
| 25 | −5031.153 | −1218.448 | 5427.051 |
| 26 | −4571.611 | −3321.471 | 4931.349 |
| 27 | −2713.526 | −4408.389 | 5427.051 |
| 28 | −395.8985 | −5161.433 | 5427.051 |
| 29 | 1746.301 | −5374.252 | 4931.349 |
| 30 | 3354.102 | −3942.984 | 5427.349 |
| 31 | 4786.475 | −1971.492 | 5427.051 |

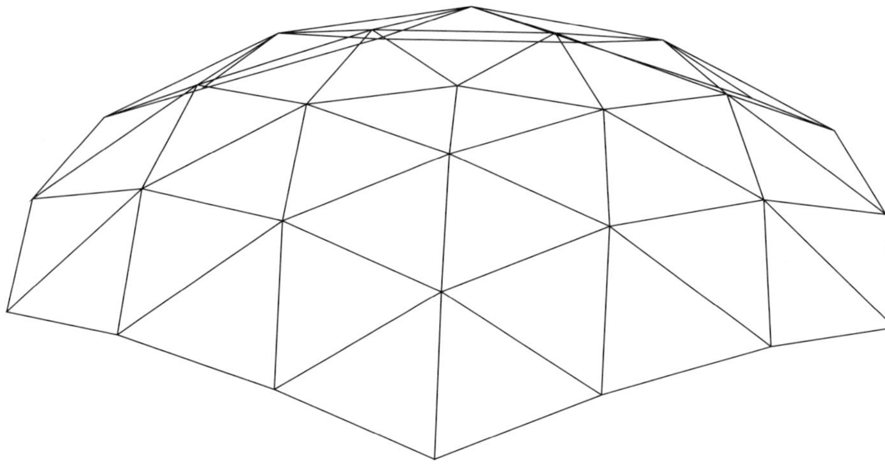

*Fig. 8.13. Perspective view of the dome*

*The number of different member lengths*

The number of members of different lengths $d_m$ is computed by using the appropriate rule [8.5].

For frequencies of $3n$, where $n = 1, 2, 3, \ldots$, and $f = 3, 6, 9, \ldots$,

$$d_m = \frac{2f^2 + 4f + 3 - 3(-1)^f}{12} \tag{8.11}$$

For frequencies of $3n - 1$, where $n = 1, 2, 3, \ldots$, and $f = 2, 5, 8, \ldots$,

$$d_m = \frac{2f^2 + 4f + 1 - (-1)^f}{8} \tag{8.12}$$

For frequencies of $3n-2$, where $n = 1, 2, 3, \ldots$, and $f = 4,7,10, \ldots$,

$$d_m = \frac{2f^2 + 16f + 51 - 3(-1)^f}{24} \tag{8.13}$$

For the dome in the present example $f = 4$ and $d_m = 6$.

Kenner [8.4] has proposed a different procedure for finding the number of different lengths involved. Referring to Fig. 8.14, the triangular face is divided into six right-angled triangles by the dotted lines OM, LE and NC. These are designated as symmetrical triangles. Consider one of these triangles OSC and count the number of members that lie wholly or partially within this triangle. Their number is 6. Hence, the number of different lengths is 6. The inventory of different lengths is given in Table 8.4.

## 8.14. Joints and roof cladding

In 1968 the first author and his colleagues at the Structural Engineering Research Centre, Chennai, India, designed and built a hemispherical geodesic dome of

12.2 m diameter and frequency 4 using class I alternative subdivision and method 2 breakdown at the Trade and Industries Fair held in the city [8.6]. The dome members consisted of 51 SWP aluminium tubes of 25–35 mm diameter (Fig. 8.15). The total weight of the fabricated structure was only 580 kg. The joints used are shown in Fig. 8.16. The triangular faces of the dome were covered with fibre-glass-reinforced polyester panels of funicular shape [8.7]. A photograph of the finished dome is shown in Fig. 8.17.

## 8.15. Stressed-skin domes

There are many advantages to be gained if the roof cladding of a dome is made to participate in the structural action. Richter [8.8] has exploited this principle in the development of Temcor polyframe domes. It is claimed that such domes are inherently more capable of withstanding heavy wind loads and drifting snow. Several examples of such domes can be found in Richter [8.8].

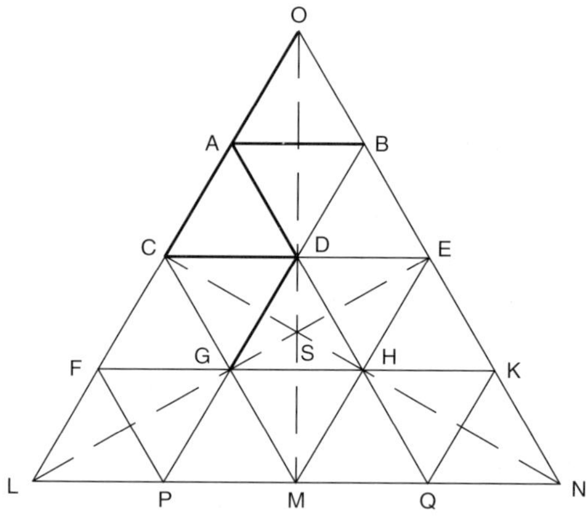


*Fig. 8.14. Kenner's method for calculating the number of different lengths of the members*

*Table 8.4. The different lengths of the members*

| Member | End | $\theta$ (rad) | $\phi$ (rad) | End | $\theta$ (rad) | $\phi$ (rad) | Chord factor | Length (mm) |
|---|---|---|---|---|---|---|---|---|
| OA | O | 0.0000000 | 0.000000 | A | 0.2538657 | 0.0000000 | 0.2531845 | 1898.884 |
| AB | A | 0.2538657 | 0.000000 | B | 0.2538657 | 1.2563700 | 0.2951875 | 2213.906 |
| AD | A | 0.2538657 | 0.000000 | D | 0.4636476 | 0.6283185 | 0.2945310 | 2208.982 |
| AC | A | 0.2538657 | 0.000000 | C | 0.5535744 | 0.0000000 | 0.2985883 | 2239.412 |
| CD | C | 0.5535744 | 0.000000 | D | 0.4636476 | 0.6283185 | 0.3128690 | 2346.518 |
| DG | D | 0.4636476 | 0.628319 | G | 0.7617826 | 0.3907125 | 0.3249199 | 2436.899 |

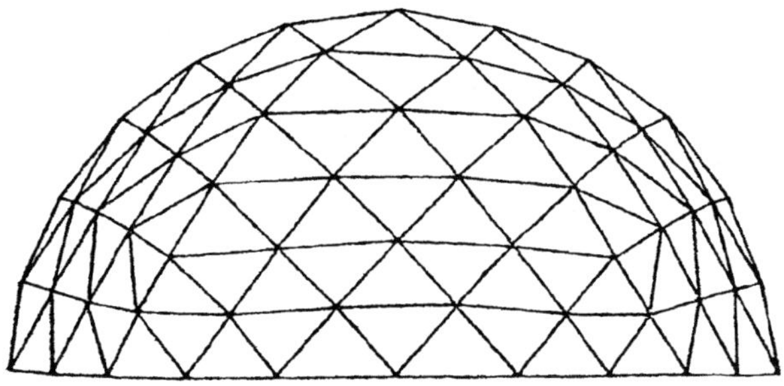

*Fig. 8.15. A hemispherical dome of frequency 4*

## 8.16. Double-layer domes

For spans in excess of 40 m, double-layer domes are generally more economical. Two examples of such domes are discussed in Sections 8.17 and 8.18.

## 8.17. Dome of the Negera National Indoor Stadium, Kuala Lumpur, Malaysia

The plans of the top and bottom layers and the diagonal bracings of the dome are shown in Fig. 8.18. Built in 1984 using the MERO system, the dome has a diameter of 91.592 m and a rise of 15 m. The top layer has 15 triangulated rings with subdivisions progressively increasing with the increase in diameter from the top to the bottom. The increases in the subdivisions are made at transition zones. The top layer is mapped on a sphere of 77.177 m radius. The bottom layer, formed on a concentric sphere of 76.617 m radius, is partially triangulated. The configuration of the bracing diagonals presents a honeycomb pattern. The bottom ring rests on alternate nodes, which provide 48 support points. The inventory of tubular members and nodes is presented in Table 8.5.

The total weight of the dome is 167.1 t. This translates into a steel consumption of 25.3 kg/m$^2$ for the projected ground plan and 22.9 kg/m$^2$ for the curved surface. Buckminster Fuller once observed that the consumption of steel in kilograms per square metre is a measure of the structural efficiency of a roofing scheme. By this yardstick the Kuala Lumpur dome may be rated as a very economical solution for covering a projected area of 6600 m$^2$. The completed dome is shown in Fig. 8.19.

## 8.18. Case study: the design, production and realization of a 50 m dome in Bremen, Germany

This case study, contributed by Prof. Dr Ir. Eekhout, highlights the problems that arise when a structural engineer has to work within the constraints set by a frozen architectural global plan and contractual compulsions oblige him to opt for a structure with less than the desired degree of reserve strength. The research and development that had to be undertaken to arrive at a suitable protective treatment for the steel components exposed to the highly corrosive environment of a swimming

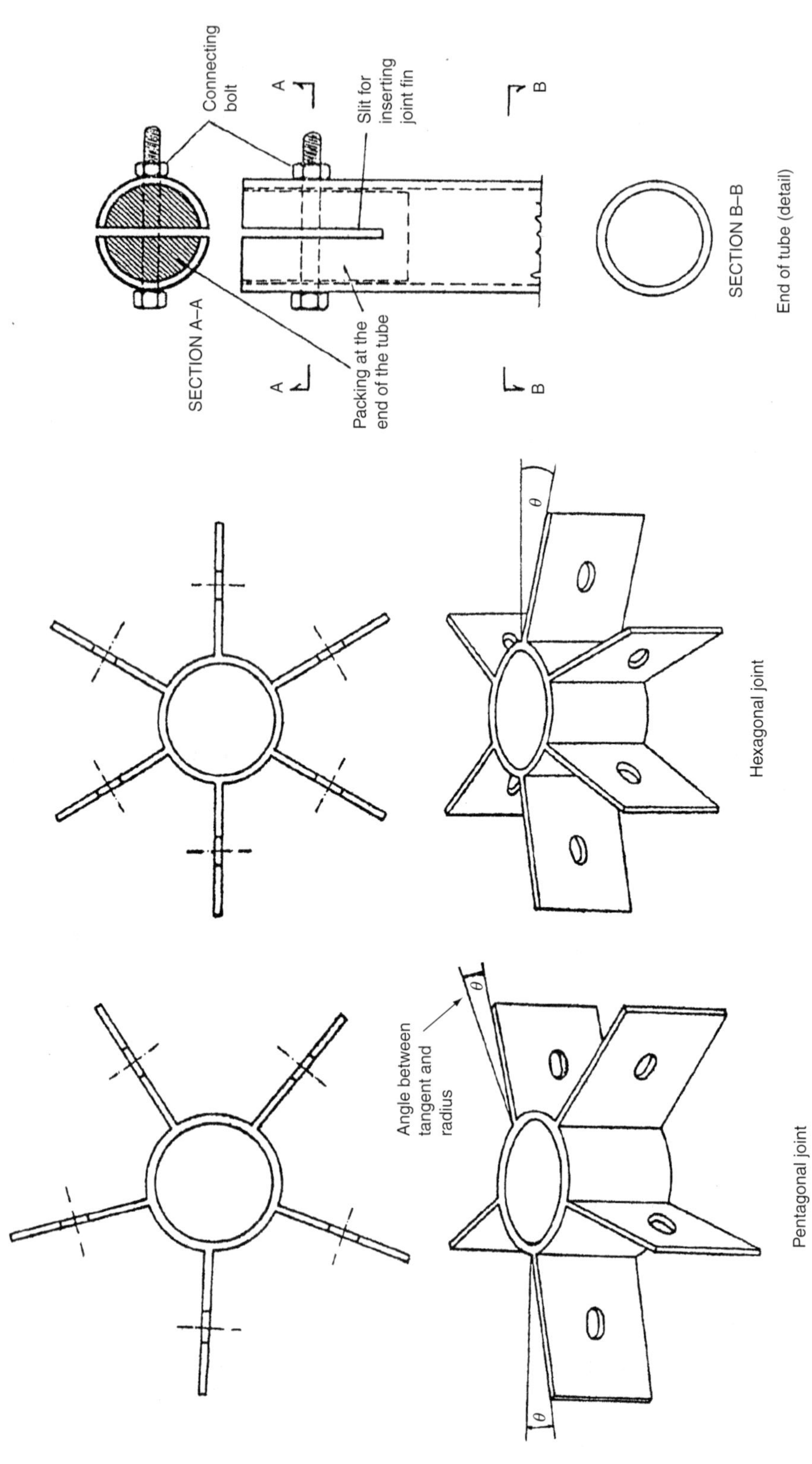


*Fig. 8.16. The joints used in the hemispherical dome of frequency 4*

*Fig. 8.17. The finished dome at the Trade and Industries Fair 1968*

pool is documented in some detail. The case study concludes by referring to a minor lapse in quality control that led to a near disaster during construction.

### *8.18.1. Background*

In the late 1980s a Dutch developer, specializing in sports complexes, undertook the development of the Aquadrom, a subtropical swimming pool at Bremen, a city in the north of Germany. The 50 m dome formed a part of complex comprising a large glass covered swimming pool with artificial waves and a smaller sauna area with smaller pools covered by a smaller dome, a restaurant area and dressing cabins. The complex was designed by the Dutch architect Karel L. Hamming, Soest. Because of the often rainy weather and the lack of sunny days, it was proposed to cover the swimming pool with a giant glass dome. The original design consisted of 12 large laminated timber arches supporting purlins covered by transparent pneumatic cushions. Because of the apprehension that the ambient humidity may affect the timber laminations, the client was open to other alternatives. Hence, the project developer, Monte Mare of Germany, invited bids from Mero of Würzburg, Germany, and Octatube, Delft, The Netherlands, for glass-covered space frames as an economical alternative to timber trusses.

The dome, as conceived in the original design, consisted of 12 segments. The brief required that the sectional profile should not be altered because the local authorities had already approved the global form. Octatube would have preferred to design a

(a)

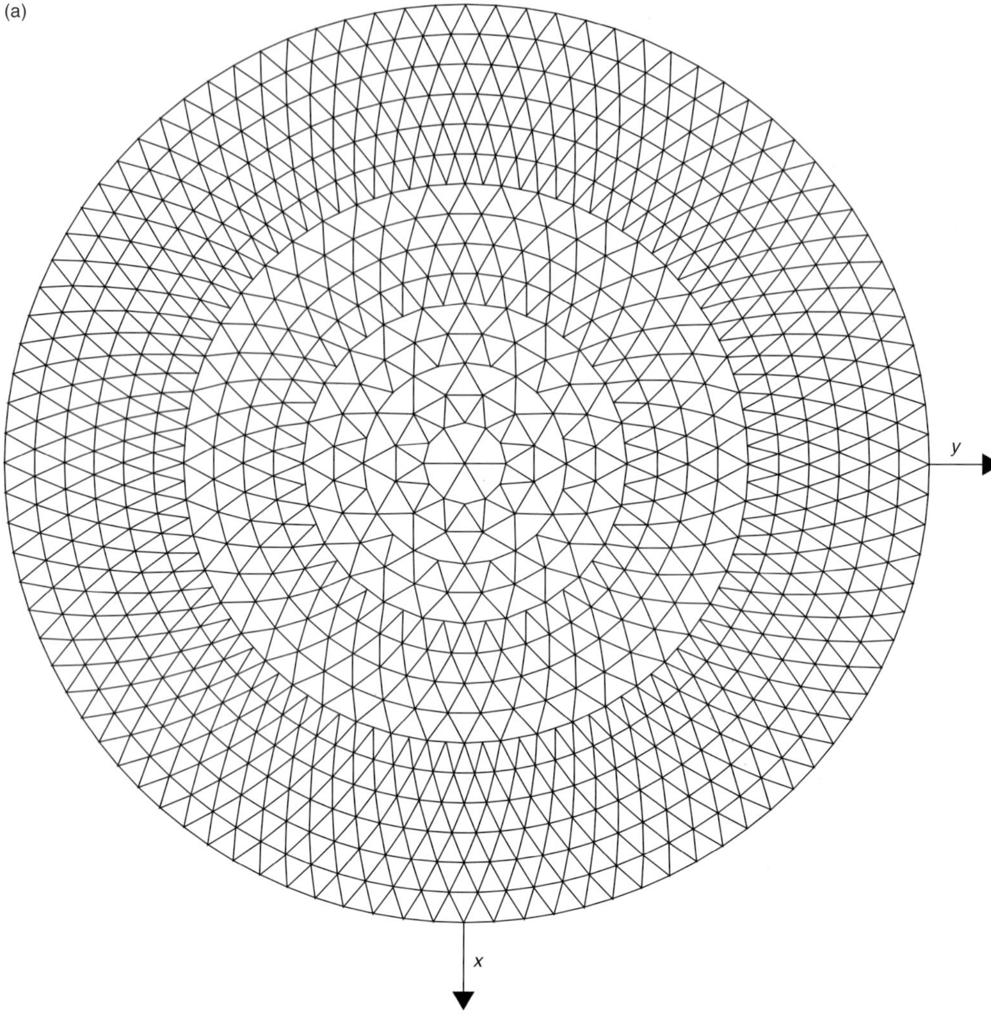


*Fig. 8.18. Plans of the dome of the Negara National Indoor Stadium, Kuala Lumpur: (a) top layer*

regular dome. To develop the design within the imposed constraints called for an enormous amount of design effort. It was quite obvious from the very beginning that to design a steel space frame to conform to the geometry dictated by the profile of timber trusses was a mistake. However, because of the obduracy of the client, Octatube was left with no other choice.

### 8.18.2. Evolution of the alternative design

Initially, Octatube proposed a dome in the form of a full double-layered space frame, with the characteristic features of progressively reduced widths and a transition to reduced double module sizes as one proceeds to the apex. The tube members in this design turned out to be slender, with sizes ranging from 60 mm × 3 mm to 90 mm × 4 mm. The numerous members at the top of the dome would produce a

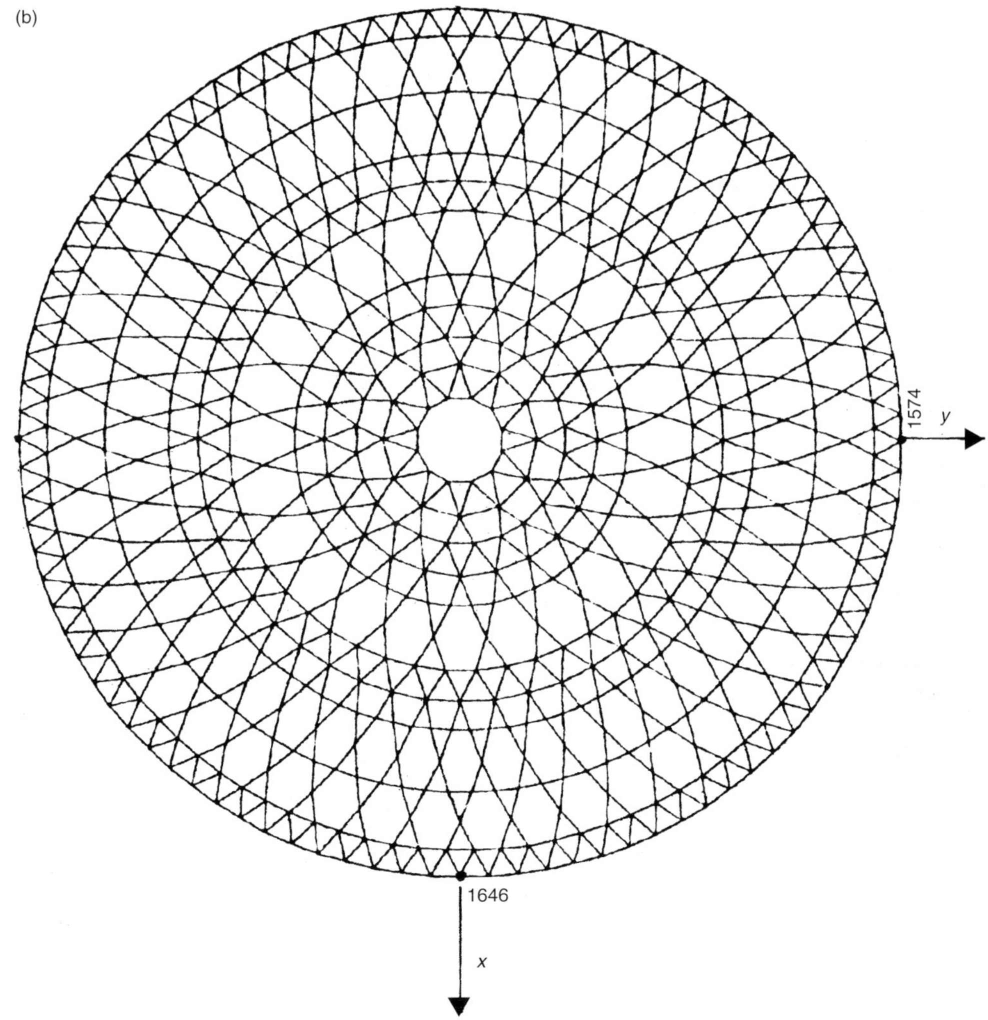


*Fig. 8.18 (Contd). (b) Bottom layer*

dazzling effect. The dome had more rigidity than required, and therefore a reduction in the number of tubular members was logical. The result was a hollowing of the geometry by reducing the number of members as one proceeded to the top. A number of wire models were made and compared to arrive at the best solution.

The dome was to be covered with insulated glass panels, with a specially designed aluminium skylight system with internal drainage. The glass panels of size 1 m × 3 m would be supported on all sides by aluminium profiles. A system of horizontal rectangular hollow sections would act as purlins to support the longer aluminium profiles at 3 m intervals. A length of 3 m is a suitable size for glass panels (jumbo float glass, 6 m × 3.21 m, can be cut to the required size, 3 m × 1.07 m, with little waste of material). The 3 m length also suited the aluminium mullions and the steel purlins of the space frames, which are commercially produced in 6 m lengths.

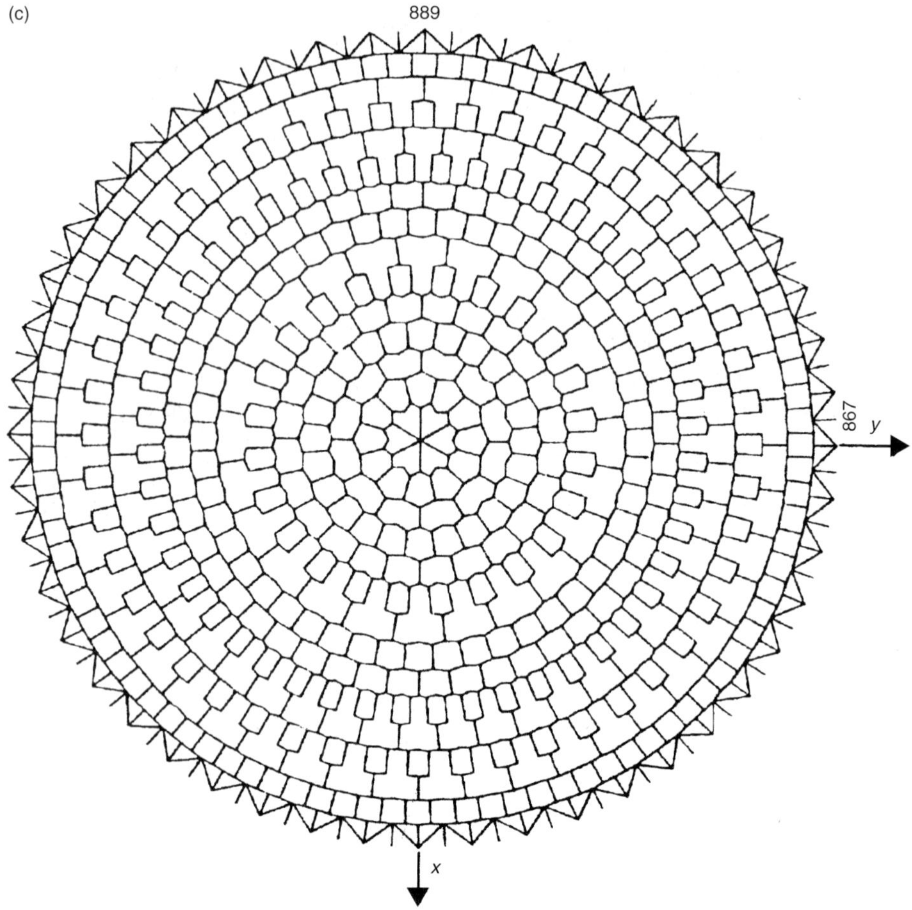


*Fig. 8.18 (Contd). (c) Diagonal bracings*

In the customary way of designing space frames, it is the Octatube practice to effect a transition from full square on square to square on double square topology, wherever possible, in order to reduce the number of members and node connectors. In general, for a perimeter-supported dome, a full square on square configuration provides more rigidity than required. This excess rigidity can be usefully exploited only if there is a large span, shallow depth or cantilevers, especially if the cantilevering is all around the structure. By following this strategy, Octatube had been able to produce structures that are competitive in price in The Netherlands with traditional steel structures.

### 8.18.3. The change to delta trusses

The client's request for a reduction in the overall price of the roof structure during the negotiating phase of the contract led to a radical change in the design concept. To reduce the cost it was necessary to reduce the number of tubular members and nodes. The dome concept had to be abandoned in favour of intersecting

(d)

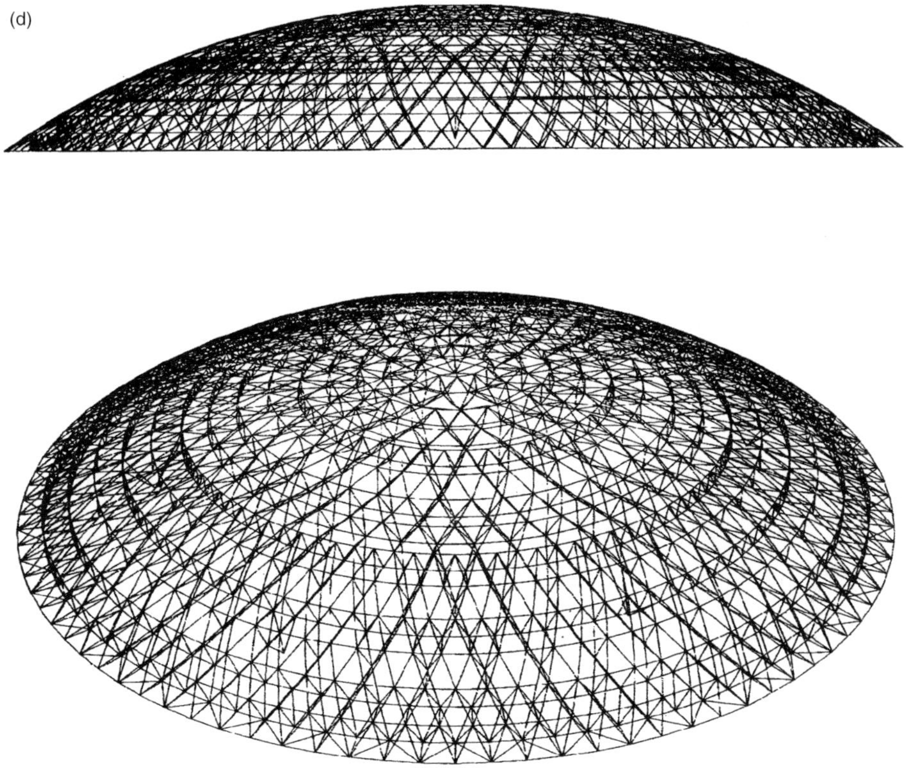

*Fig. 8.18 (Contd). (d) Perspective view*

*Table 8.5. Inventory of the tubular members and nodes used in the dome of the Negera National Indoor Stadium*

| | Number | Diameter (mm) | Wall thickness (mm) | Length (m) |
|---|---|---|---|---|
| *Nodes* | | | | |
| Top | 937 | 87 and 100 | – | – |
| Bottom | 721 | Up to 178 and 198 | – | – |
| Total | 1658 (19.1 t) | | | |
| *Members* | | | | |
| Top | 2622 | 60.3 to 159 | 2.9 to 6.3 | 1.8 to 3.6 |
| Bottom | 1607 | 60.3 to 159 | 2.9 to 6.3 | 1.8 to 3.6 |
| Diagonal | 2273 | 60.3 to 159 | 2.9 to 6.3 | 1.8 to 3.6 |
| Total | 6502 (148.0 t) | | | |

independent delta trusses, separated by open spaces, with purlins spanning over more than 9 m to bridge the gaps. The dome was substituted by a 12-segment agglomeration of upwardly curved trapezoidal modules. Each module consisted of two radial trusses on the two radial sides, interwoven with delta trusses at the top and bottom (Fig. 18.20). In replacing the dome by delta trusses the most important attribute of a monolithic space frame – its high degree of static indeterminacy – was lost. This reduced the possibility for internal forces to take other routes for load transfer in the event of eventual collapse of the dome due to breaking of bolts or nodes, or yielding of bolts, nodes or tubular members.

The local failure of a spherical node, during installation at Bremen, highlighted the importance of indeterminacy. The reduction in the number of tube members to a minimum to accommodate the price cut and the consequent loss of indeterminacy made the dome more vulnerable to the failure of a single element.

### 8.18.4. Structural analysis and design

Deviating from the normal design practice, the space frames were regarded as partially fixed.

*Fig. 8.19. The completed dome of the Negera National Indoor Stadium, Kuala Lumpur*

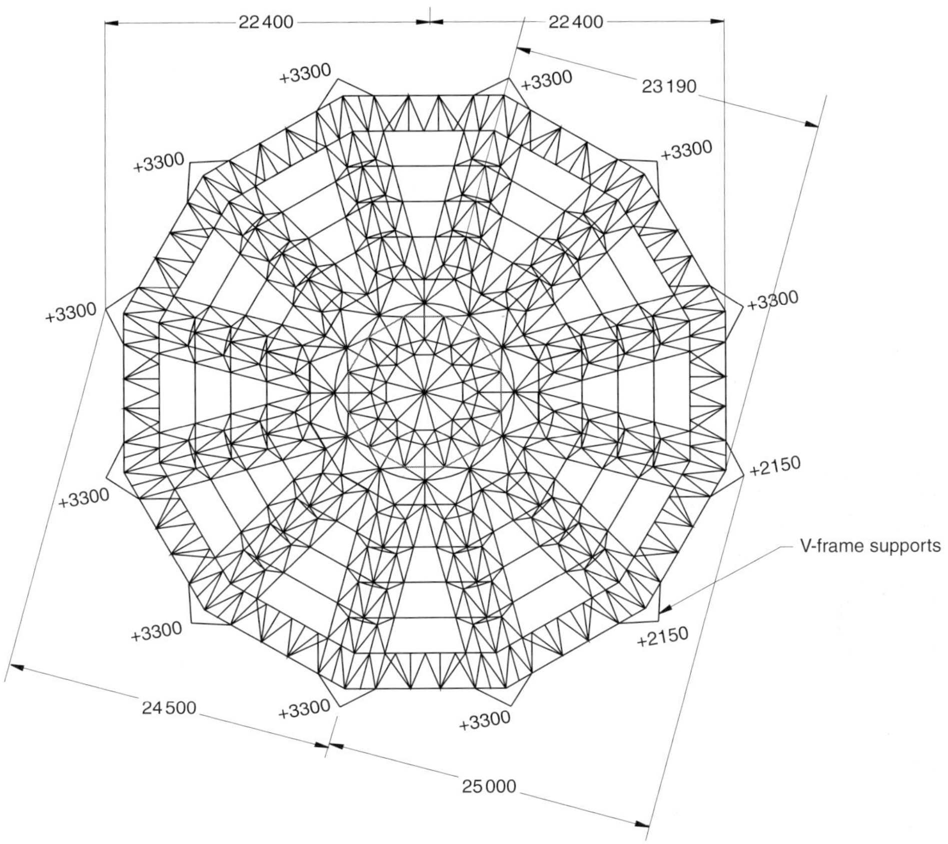


*Fig. 8.20. Plan of the dome composed of delta trusses*

The snow load of 0.75 kN/m$^2$ was assumed to act on parts of the dome where the slope is shallower than 45°. The wind load was assumed to be 0.8 kN/m$^2$. With the exception of the lower windward side of the dome, which would be subject to compression, the rest of the dome would be subject to varying degrees of suction. The governing loads would be symmetrical and asymmetrical combinations of dead load, live load (0.4 kN/m$^2$), wind load and snow load. Another load considered was the weight of a maintenance ladder at the top of the dome, although after the engineering of this ladder was completed the client did not venture to have this 25 m long rotating bridge ladder produced and installed.

Analysis was carried out using the SAP90 software. The tube sizes resulting from the analysis and design ranged from 60.3 mm × 2.9 mm to 101.6 mm × 5 mm for the upper chord members, from 60.3 mm × 2.9 mm to 82.5 mm × 3.2 mm for the diagonals and from 60.3 mm × 2.9 mm to 88.9 mm × 3.2 mm for the lower chord members. The resulting dead weight was less than 0.18 kN/m$^2$.

The V-framed pairs of circular hollow section columns, 114.5 mm × 12.5 mm and 127 mm × 12.5 mm in size, received high concentrated forces. The footings of the columns were detailed with four anchors and a solid stud under the base plate. The footings were placed in a concrete void to be filled in later in order to provide adequate shear resistance.

### 8.18.5. Node connectors

The Tuball spherical node connectors of 162 mm diameter were tested at TNO under a biaxial state of stress with tensile and compressive forces applied at right angles. The maximum breaking load was found to be 800 kN. This strength was sufficient at all locations except over the V-columns where, including a safety factor of 1.5, a node with a breaking strength of 1800 kN was required. At these locations a solid node was used. The use of solid spheres is a deviation from the normal Tuball system and it called for a particular sequence of bolting. The tubes were first connected to a solid sphere and then to the next hollow sphere, which provided easy accessibility. Similarly, the base plates had to be screwed first to the solid nodes before they were installed over the cast-in anchors on the top of the V-columns.

### 8.18.6. Bolts

The connecting axial high-strength grade 8.8 bolts were of M24, M20, M16 and M12 size. The high forces transmitted through the solid spheres to the columns called for grade 8.8 bolts of size M27.

### 8.18.7. Dome profile and column heights

The dome, as finally built, had a rise of 7.5 m, which is much less than the optimum value. The dome initially proposed by Octatube had a rise of 12.5 m above the V-framed columns.

### 8.18.8. Small dome and entrance skylight

The smaller sauna dome with a span of 12 m was a conventional structure with rectangular hollow section profiles acting as radial legs and the skylight system as the radial and meridianal connecting elements. These elements acted both structurally and functionally, and also formed watertight supports for the insulated glass panels. This dome was completely assembled, including the skylight elements, at the ground level and placed in position on the roof in one erection operation by means of a 250-t mobile crane.

The entrance skylight was a semi-cylindrical dome. It was designed as a horizontally curved space frame in the form of triangular trusses with almost square insulated glass panels. Because of the smaller span, the outer space frame members were made of extruded aluminium. The spherical nodes were made of cast aluminium. All aluminium and galvanized steel elements were powder coated. The front facades were designed in a radial fashion.

### 8.18.9. Assembly and erection

The space frame segments were assembled on steel stilts at ground level. The lower chord members and nodes were assembled first, followed by assembly of the

*Fig. 8.21. A segment of the dome during hoisting*

diagonal members. Next, the upper chords and nodes were added. These assembled elements and components were tightened by means of torque wrenches and the node caps were then closed. The purlins were also assembled at ground level, as were the aluminium mullions and transoms, including the internal water draining system. The assembled space frame and skylight mullion segments were erected by means of a 250-t mobile crane in three hoisting operations involving four segments each (Fig. 8.21). The segments were supported on the outer V-columns and at the centre of the dome on a temporary space frame tower erected to support the top ends of the space frame segments (Fig. 8.22). The foot of the tower was equipped with hydraulic jacks.

The first segments, even numbers 10 and 11, seemed to fit exactly during the erection operations. Of course, the whole engineering process of design, analysis, production, assembly, installation and erection of an innovative structure is like developing a prototype, with the unavoidable risk of human error at each step. Although Octatube had at that time a 10-year track record of developing experimental structures, each phase could still produce an unforeseen flaw, ending in disaster. Would the last segment, number 12, fit between segments 1 and 11? The details did not allow a tolerance greater than 8 mm. Hoisting the last segment took place in an atmosphere of high drama, akin to that witnessed at the launching of an American rocket from Cape Kennedy. Would the segment slide in, or would it be, for example, 20 mm too wide? And what to do if this happened? Fortunately, or due to the engineering and production skills of the Octatube crew members, the last

*Fig. 8.22. The temporary space frame tower support*

module fitted like a glove. A distance of 2–4 mm was left between the various adjacent nodes, which could easily be closed by tensioning the M12 bolts between the two circular steel plates containing four bolt holes. The last two stitching lines of the nodal sides of module 12 distributed the few millimetres of space over two lines, and these are closed by tightening over more stitching lines adjacent to the other adjacent sides.

It is indeed remarkable that the last segment fitted almost exactly between the previously erected 11 segments, although the team responsible for this project had no previous experience of designing or building a similar dome. The completed dome is shown in Fig. 8.23. Was the success due to good luck or to competent engineering? Good luck almost does not exist in engineering, so the success was the result of meticulous, methodically ordered design and engineering, with its obligatory external and voluntary internal checking systems. Remarkably, only a few projects in the history of Octatube have been marred by unexpected or expected major problems that could not be overcome. Only one project in the course of its 20 years of experience could not be built. This was a glass beam of 27 m span to be included in the Dutch head office of the life insurance company Zwitserleven in Amstelveen, The Netherlands. The office was to be near Amsterdam Airport, and was designed by the architect Pi de Bruin. Detailed studies of this project, followed by extensive practical tests on different scales, did not lead to reliable results, despite the calculations of the external project engineer. So, at the end of a 2-year structural development process, the entire budget had

*Fig. 8.23. The Aquadrom, Bremen, Germany: (a) the dome near completion; (b) the finished dome*

been spent on development. The lesson learnt is that fundamental developmental work should not be combined with a project under execution. The required and available time frames are not compatible.

### *8.18.10. Corrosion protection*

A matter of great concern to the local authorities was the chemically aggressive environment of the swimming pool. The apprehensions about corrosion stemmed from a report published some time earlier on the collapse of a swimming pool ceiling in Switzerland because of corroded stainless steel suspenders. The usually superior quality of stainless steel was apparently not good enough for a swimming pool environment. For this reason, extensive chemical tests were performed at the Faculty of Metallurgy of the University of Technology at Delft, which is located opposite Octatube's office in Delft. For Octatube the close proximity to Delft University of Technology has proved to be very beneficial, providing an excellent environment for interaction between theory and practice.

The results of the material testing produced some surprises in the choice of materials for the protective coatings. Comparative tests of different types of stainless and galvanized steels, both pure and with an extra top layer of powder coating, showed that the best results were achieved by hot dip galvanized, powder-coated steel. Stainless steel did not show a non-corrosive response, and in general the chloride atmosphere was highly aggressive to metals. Usually stainless steels are seen as superior to mild steel and high yield steels. However, the mild steel protected by a hot dip galvanized layer followed by powder coating proved to be quite resistant to the swimming pool environment. Based on these findings, all steel components were hot dip galvanized after welding in the shop and were subsequently powder coated. Spraying circular elements with paint does not entirely and uniformly cover their surface, but a powder coating will be evenly distributed, because of the electrostatic distribution of the powder particles all over the envelope of the sphere and the tube. Powder coating thus produces a protective layer of uniform thickness. The bolts were also all hot dip galvanized and powder coated. Such double protective treatment is rare for hidden bolts, but in this case the test results indicated that it was necessary. The cap of the node was subsequently glued onto the top of the shell part of the node in order to restrict the amount of active chloride-rich air that could enter. Tubular members were not sealed off with silicone, as the volume of air trapped inside the node was small. When pressed up to the spheres the tubes are quite closed, and the resulting combination is not likely to suffer corrosion. In the 10 years that have elapsed since the inauguration of the swimming pool, which has been in continuous use, no reports have been received of any damage caused by corrosion.

### *8.18.11. Cracking of a cast node*

In the Bremen project, a minor detail caused a major problem. After the completion of the installation, the hydraulic jacks of the temporary tower were relieved, and instantaneously the 25 m span became a 50 m span. At that moment, the Chief Engineer of Octatube, who was present at the site, heard a crack during the jack release, which is a rare experience for a structural engineer. It appeared that one of the spheroidal graphite nodes in one of the lower tensile rings of the dome had cracked. This crack ran vertically between the large holes and the top edge of the sphere. The immediate reaction of the engineers was to raise the tower again by pumping up the jacks and to start a comprehensive process of investigation to diagnose the cause of the crack. No advantage was taken of the topological reserve strength of the dome, as the engineers believed that the cause of the crack and its possible propagation to other spheres should first be determined before relying on the overall strength of the dome.

This research process took about 4 weeks. After the initial crack, the first impulse was to replace the cracked spherical node. After a day's work, the replacement was in place and the tower was lowered again. Simultaneously, an inspection of the nodes was initiated. After some days of investigation it appeared that the process of casting the spheres had led to the problem. The node was produced in a grey iron casting. It was discovered that, at the end of the casting of a series of spherical elements, the residue at the bottom of the casting pan, which contains more carbon than the normal mix, was also poured into the sand moulds, resulting in an extraordinarily carbon-rich mix. Metallurgical analysis of the cracked node done at the University of Technology at Delft established the high carbon content of the cracked node. Answers to the following questions had to be found:

- How many spherical nodes with a high carbon content had gone into vulnerable locations such as the lower horizontal rings?
- How dangerous were the consequences?

To find the answers to these questions the following actions were initiated:

- an analysis of the production process
- an analysis of the quality assurance schemes
- the determination of the exact number of high carbon content nodes
- the replacement of the built-in high carbon content nodes
- an improvement of the quality assurance procedures.

All spherical nodes in the built structure had to be investigated for possible cracks, which could not found by visual inspection. Although an acoustical inspection would have provided a good basis for assessing individual nodes, electro-sonar control inspection would have been even better. However, the expense and effort involved made the latter virtually unaffordable. The engineers at Octatube had to find a simpler but suitable method for checking the quality of the nodes. The solution was found in the drilling of the spheres.

The raw spherical nodes received from the foundry had been machined until a smooth surface with the desired outside diameter was obtained. Later, the bolt holes were drilled. Confronted by the unexpected rupture of a sphere, the head of the machining department remarked that he had noticed that some of the spheres had produced drilling dust instead of the normal steel curls during machining, but at that time he had not given the matter much attention. Thus all the spherical nodes installed in the dome were drilled. None of the spheres produced any drilling waste other than a firm curl, which proved that all the built in spherical nodes had the desired composition. However, two spheres that were not built in, as they were spare parts, were detected which produced a dull sound instead of a clear ring when struck, and on drilling these nodes it was found that they had to be rejected.

The whole episode showed that the manufacturer of the spherical nodes did not perform a thorough exit quality control and that the entrance quality control performed at Octatube was not quite adequate. As a remedial step a new regime of quality assurance, both at the foundry and at Octatube, was instituted. Publication of this narrow escape seemed worthwhile in order to underline the dictum of the famous American architect Ludwig Mies van der Rohe, that 'God is in the details'. Or, as in this instance, 'The devil is in the details', a common German saying ('Der Teufel steckt im Detail').

### *8.18.12. German Prüfungsingenieur system*

The Bremen project involved the German system of the Zulassungsingenieur and Prüfungsingenieur. Octatube regarded these as an added burden on top of the continuous responsibilities and liabilities of Octatube as the prime engineering and producing company. It resulted in an overcautious approach resulting from dual checks to eliminate any risk by the two controlling and demanding parties. The checking system ended in some elements and components that were heavier and more costly than is usual in normal Dutch practice. The Prüfungsingenieur checked all computer calculations by hand calculations (nowadays the computer analysis would be checked by means of computer programs of another brand). All details were checked and discussed with the Octatube Engineer, Evert de Jong, and his colleagues. After submission of additional information, approval was finally received in parts. The responsibility of the Prüfungsingenieur is always to the city, while the responsibilities of the engineering firm are always to the client. (By the way, the client went bankrupt before final completion of the building, leaving invoices unpaid, and hence the liabilities of Octatube were never activated.) The German system of checks and double checks leads to conservative designs and more expensive buildings. Only seldom does a Dutch design and build entrepreneur long to work under this system – the project involving the 27 m span glass beam would have been aborted much sooner if no Prüfungsingenieur had been involved.

## 8.19. References


8.1. MAKOWSKI, Z. S. History of the development of braced domes. *Bulletin of the International Association for Shell and Spacial Structures*, 1989, **30**(3), 169–183.

8.2. RICHTER, D. L. Space structures development from early concepts to Temcor geodesic domes. *Proceedings of the 2nd International Conference on Space Structures*, University of Survey, Guildford, 1975.

8.3. UNITED STATES PATENT OFFICE. US Patent 2682235, 29 June 1954 [relating to a framework for enclosing space].

8.4. KENNER, H. *Geodesic Math and how to Use it*, University of California Press, Berkley, CA, 1976.

8.5. HOLLINGS, J. *Geodesic Domes – The Hard Facts*, Symatrix, Wellington, 1988.

8.6. BALLAL, B.Y, ABDUL, K. E, RAMAIAH, M. and RAMASWAMY, G. S. A new type of braced dome with minimum number of different members. *Journal of Structural Engineering*, 1987, **5**(1), 1–6.

8.7. RAMASWAMY, G. S. *Design and Construction of Concrete Shell Roofs*, McGraw Hill, New York, 1968, chap. 18, pp. 464–471.
8.8. RICHTER, D. L. Developments in Temcor aluminium domes. In: Z. S. Makowski (ed.), *Analysis, Design and Construction of Braced Domes*, Granada, London, 1984, chap. 18, pp. 521–40.

# 9. Optimization techniques

## 9.1. The design process

The design process for a steel space frame involves three distinct phases:

- planning and conceptual design
- preliminary design
- final design and optimization to minimum weight.

During the preliminary design phase, one of the objectives is to arrive at the member sizes required to start the first cycle of the stress analysis. An equally important goal at this point is to determine fairly accurately the weight of the space frame, because preliminary designs form the basis for competitive bidding and weights are often decisive in winning contracts. In the archives of any large design bureau, a large volume of data on the weights of similar space frames designed in the past will normally be available. These data can be put to good use in predicting the weight of space frames at the preliminary design phase by using *artificial neural networks* (ANNs).

With the advent of computers and the availability of powerful software packages, structural analysis has become a more or less totally automated process that can be carried out even by a novice with little or no design experience. However, design still needs the input of experienced experts. *Expert systems* offer a means of transferring the rich heuristic knowledge and decision-making skills of experts to analysts who, while lacking design experience, have the needed computational skills.

Recent advances in *genetic algorithms* (GA) [9.1–9.3] and *evolution strategies* [9.4, 9.5], which mimic biological evolution, based on the Darwinian theory of the survival of the fittest, have added new and powerful tools to the armoury of the designer for optimizing the final design of space frames to minimum weight. The computational efficiency of evolution strategy methods can be further enhanced by combining them with ANN techniques. The purpose of this chapter is to introduce the reader to some of these recent developments.

## 9.2. Artificial neural networks

The unsurpassed processing capabilities of the human brain are attributable to the presence of $10^{11}$ neurons and their complex interconnectivity. The human brain represents the power of parallel processing at its best. ANNs are highly simplified models that mimic the biological neural network of the human brain.

A biological neuron (Fig. 9.1) consists of a central nucleus or cell called the *soma* to which are attached *axons*, *dendrites* and *synapses*. Dendrites receive input signals from the sensory organs and other neurons. The output from the cell is transmitted through the axon to synapses, where the outputs are multiplied by the synoptic weights before transmission to other neurons. The close similarities between the biological neuron and ANNs is apparent from Fig. 9.2, which represents the processing elements of a neural network.

### *9.2.1. The feed-forward operation*

Let the inputs (Fig. 9.3) be $x_{\mathrm{p}1}, x_{\mathrm{p}2}, \ldots, x_{\mathrm{p}i}$ and let $w_{ji}$ be their corresponding weighting functions. The net input arriving at a typical hidden node $j$ is

$$\mathrm{net}^{\mathrm{h}}_{\mathrm{p}j} = \sum_{i=1}^{N} w^{\mathrm{h}}_{ji} x_{\mathrm{p}i} + \theta^{\mathrm{h}}_{j} \tag{9.1}$$

where the superscript h denotes the hidden layer. The first term represents the sum of the weighted inputs, and $\theta_i^{\mathrm{h}}$ is the *bias*, which is in the nature of a threshold. The bias may be looked upon as an additional input of 1. At the node $j$ of the hidden layer, the input $\mathrm{net}^{\mathrm{h}}_{\mathrm{p}j}$ is non-linearly transformed by a squashing function, resulting

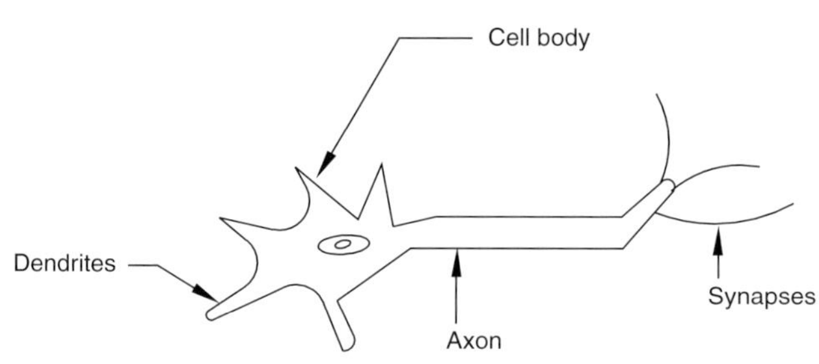


*Fig. 9.1. A neuron*

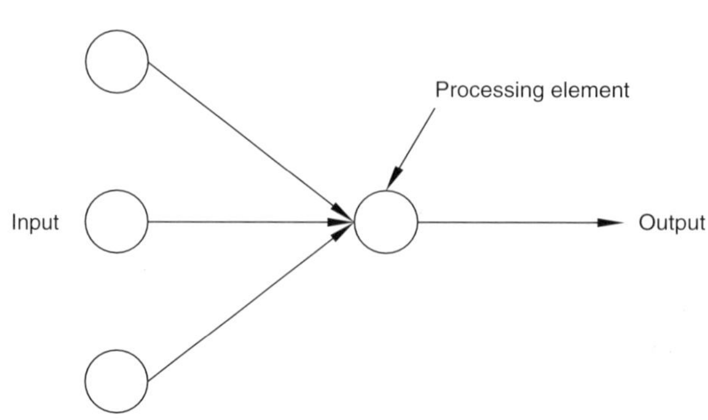


*Fig. 9.2. The processing element of an ANN*

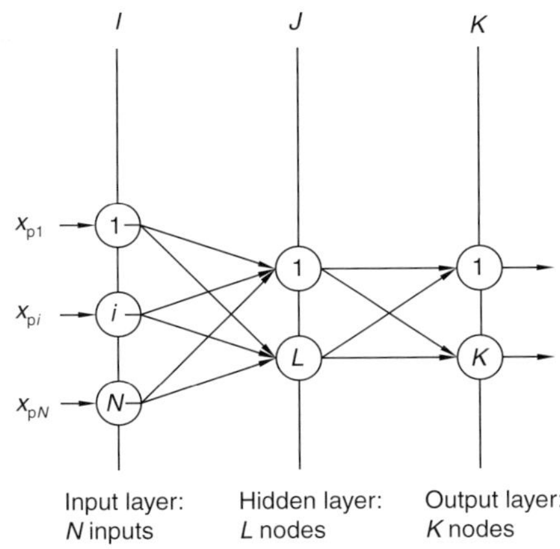


*Fig. 9.3. A multi-layer neural network*

in an output

$$i_{pj} = f_j^{h}(\text{net}_{pj}^{h}) \tag{9.2}$$

where $f_j^{h}$ is the *squashing or activation function*.

In a network with a linear processing element, the squashing function may be a simple threshold $\theta$ that generates the following outputs:

$$\text{If } \sum w_{ji} x_{pi} \leq \theta, \text{ output } i_{pj} = 0$$
$$\sum w_{ji} x_{pi} > \theta, \text{ output } i_{pj} = 1$$

In networks with non-linear processing elements the squashing function is either a *sigmoidal* or *tanh* function (Fig. 9.4). The sigmoidal or logistic function always generates outputs in the range 0 to 1. The tanh squashing function generates outputs in the range –1 to +1.

The output from a hidden node with a sigmoidal squashing function takes the form

$$i_{pj} = \frac{1}{1 + \exp[-(\sum w_{ji} x_{pi} + \theta_j^{h})]} \tag{9.3}$$

The sum of the weighted inputs arriving at an output node may be written as

$$\text{net}_{pk}^{o} = \sum_{j=1}^{L} w_{kj}^{o} i_{pj} + \theta_k^{o} \tag{9.4}$$

where the superscript o denotes an output node. The output emerging from this node after sigmoidal transformation may be written as

$$o_{pk}^{o} = \frac{1}{1 + \exp[-(\sum w_{kj}^{o} i_{pj} + \theta_k^{o})]} \tag{9.5}$$

Let the desired output be $y_{pk}$. The difference $\delta_{pk} = y_{pk} - o_{pk}$ between the desired output $y_{pk}$ and the actual net output $o_{pk}$ represents the error. The sum of the squares of the errors of all the outputs is a measure of error and is a good index of how well the net is performing. This measure is usually represented by

$$E_P = \tfrac{1}{2} \sum_{k=1}^{M} \delta_{pk}^2 \tag{9.6}$$

The factor of a half is introduced to facilitate differentiation.

What we have just described is the feed-forward operation. The errors are now back-propagated and the weights are modified. The feed-forward operation is then repeated. This cycle of forward and backward pass is continued until the errors are within acceptable limits. The net is now said to be *trained* and it can be used to generate outputs corresponding to given inputs. The progressive updating of weights by using desired outputs corresponding to each set of inputs is known as *supervised training*. The feed-forward multi-layer neural network just described with back-propagation of errors and supervised learning is generally used for all structural engineering applications.

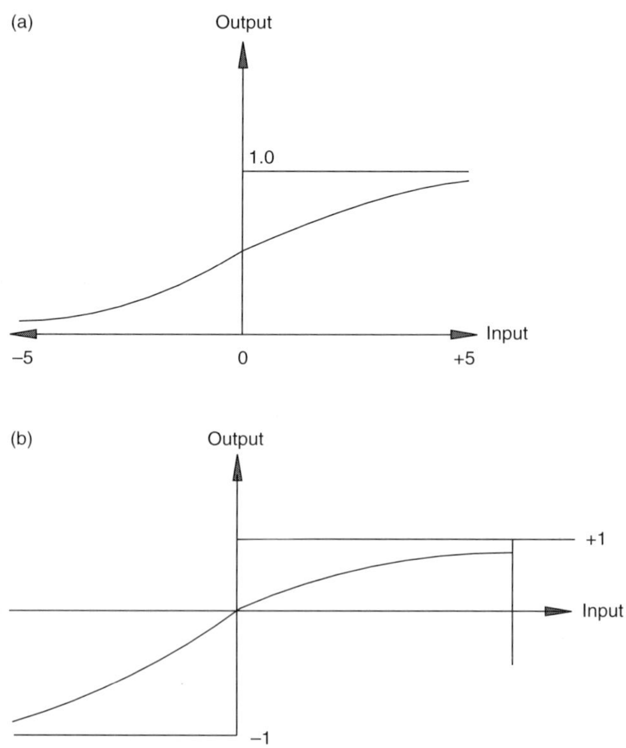


*Fig. 9.4. (a) Sigmoidal and (b) tanh activation functions*

*Table 9.1. Inputs and outputs for the XOR problem*

| Input | | Output |
|---|---|---|
| $x_1$ | $x_2$ | |
| 0 | 0 | 0 |
| 0 | 1 | 1 |
| 1 | 0 | 1 |
| 1 | 1 | 0 |

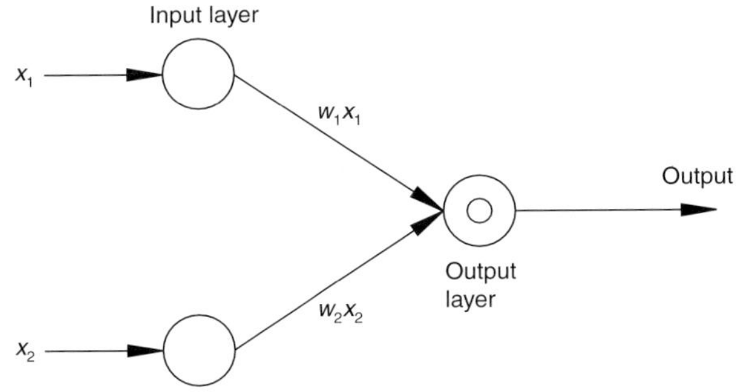


*Fig. 9.5. The inputs and outputs for the XOR problem*

### 9.2.2. Selection of network topology

The selection of the best topology involves decisions relating to the number of hidden layers and the number of nodes or processing elements in each layer.

In almost all problems of practical interest, the similarity structure of the input and output patterns will be different. Such problems cannot be solved without an internal representation (a network without hidden layers with processing units). A classic example is the *exclusive-or* (*XOR*) *problem*, the patterns of inputs and outputs for which are given in Table 9.1.

This problem cannot be solved without a hidden layer. Referring to Fig. 9.5, the problem may be stated as

$$\text{If } w_1x_1 + w_2x_2 \geq \theta, \text{ output} = 1$$
$$w_1x_1 + w_2x_2 < \theta, \text{ output} = 0$$

It is not possible to find a linear threshold $\theta$ that will satisfy these relationships. In geometrical terms this means that a hyperplane cannot be found such that the pair of inputs which give the same output lie on the same side of the plane [9.6]. This can be clearly seen from Fig. 9.6.

The XOR problem is easily solved once a hidden layer (Fig. 9.7) is introduced. The threshold values $\theta$ are shown enclosed in circles in Fig. 9.7. It may therefore be concluded that one hidden layer is absolutely necessary. It has been shown that with two hidden layers the feed-forward neural network becomes a universal mapper. It has the power to solve any problem if there are no constraints on the number of processing elements in each hidden layer or on the time required to reach a solution [9.7].

There are only heuristic rules available for deciding on the number of processing elements on the hidden layer. For a three-layer network with $k$ inputs, it is not necessary to have more than $k-1$ nodes [9.8]. In arriving at the most suitable topology, the two conflicting considerations of the speed of learning and the ability to generalize have to be reconciled. The learning time is proportional to $N^3$, where $N$ is the number of weights in the network. Ideally, the number of connections must be the minimum consistent with the ability of the net to generalize. The procedure normally followed is to start with one hidden layer and add additional layers if this is found to be necessary.

The worked example that follows illustrates the application of an ANN to a beam problem.

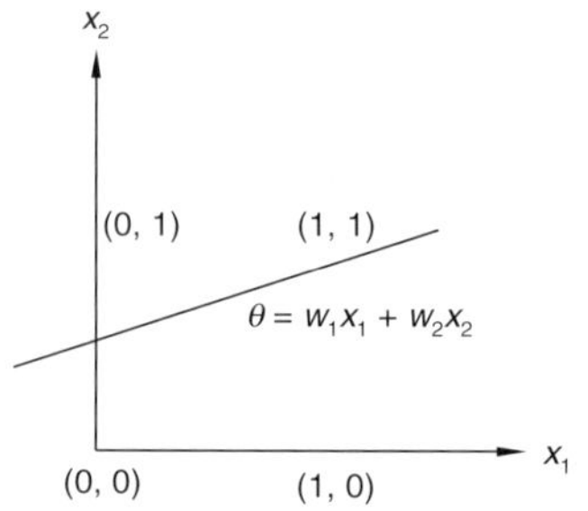


*Fig. 9.6. Graphical representation of the XOR problem. (Reproduced with permission from Freeman and Skapura [9.6])*

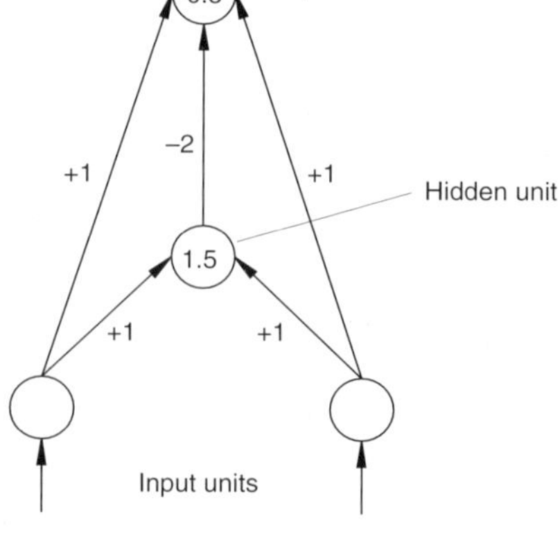


*Fig. 9.7. An XOR network with one hidden unit*

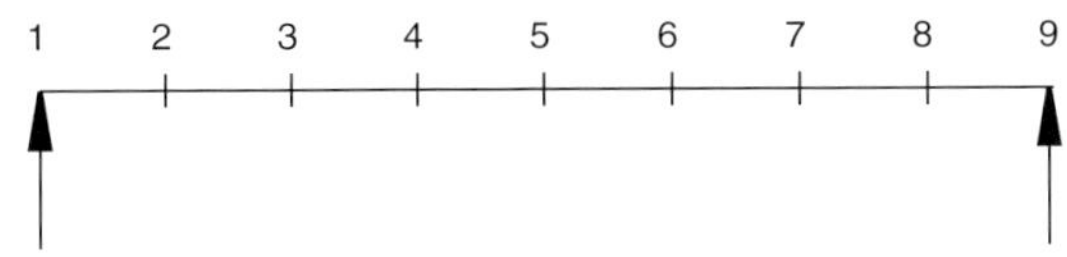


*Fig. 9.8. An ANN applied to the beam problem in Example 9.1*

*Table 9.2. The inputs and outputs for the beam problem in Example 9.1*

| Load at point | Bending moment at point | | | | | | | Deflection | | Error (%) |
|---|---|---|---|---|---|---|---|---|---|---|
| | 2 | 3 | 4 | 5 | 6 | 7 | 8 | Computed | Predicted | |
| *Training set* | | | | | | | | | | |
| 2 | 0.0875 | 0.075 | 0.0625 | 0.05 | 0.0375 | 0.025 | 0.0125 | 0.1568 | 0.1773 | 13.07 |
| 3 | 0.075 | 0.15 | 0.125 | 0.10 | 0.075 | 0.05 | 0.025 | 0.2932 | 0.2995 | 2.15 |
| 4 | 0.0625 | 0.125 | 0.1875 | 0.15 | 0.1125 | 0.075 | 0.0375 | 0.39 | 0.3988 | 2.26 |
| 5 | 0.05 | 0.100 | 0.15 | 0.2 | 0.15 | 0.10 | 0.05 | 0.4270 | 0.4297 | 0.63 |
| *Test set* | | | | | | | | | | |
| 3 and 5 | 0.125 | 0.25 | 0.275 | 0.3 | 0.225 | 0.15 | 0.075 | 0.72 | 0.684 | 5.00 |

### *Example 9.1*

The beam shown in Fig. 9.8 is equally divided into eight parts and a unit load is successively applied at points 2 to 4, generating the bending moment patterns given in Table 9.2. The computed central deflections corresponding to these bending moment patterns may also be found in Table 9.2. Regarding the bending moment patterns as inputs and the computed central deflections as the output, our aim here is to train a neural net and use the net to predict the deflection at the centre corresponding to the given test input pattern.

*Solution*. A triple-layer feed-forward and error back-propagation network that comprised one hidden layer with four processing elements on it was selected. The number of inputs including bias is 8 (Fig. 9.9).

The final weights and biases arrived at using the network are given in Table 9.3. The deflection of 0.68 predicted by the net agrees closely with the computed deflection of 0.72 for the test case with unit loads at points 3 and 5.

## *9.2.3. Predicting the weights of space trusses using an ANN*

Let us suppose that a design bureau has data relating to the weights of several space trusses designed in the past. An ANN then offers a means of predicting the weights of other trusses of the same topology with a fair degree of accuracy.

A space truss, rectangular in plan, is characterized by the following seven parameters:

Length, $l$
Breadth, $b$
Structural depth, $d$
Grid dimension, $g$
Total load acting, $P$
Average column spacing, $s$
Weight of space truss, $W$

It is advantageous to represent the input and output variables in dimensionless form using Buckingham's $\pi$ theorem [9.9]. The number of variables is $m = 7$ and they involve $n = 2$ fundamental dimensions of force and length. Hence, the number of dimensionless $\pi$ terms, according to the theorem, is $m - n = 7 - 2 = 5$. The five terms are:

Aspect ratio, $b/l$
Depth/shorter span, $d/b$
Depth/grid dimension, $d/g$
Average spacing of columns/span, $s/b$
Weight of space truss/total load, $W/P$

The spacing will be described as 'close' or 'sparse', depending upon the value of $s/b$. The procedure for predicting the weight is explained below by means of an example.

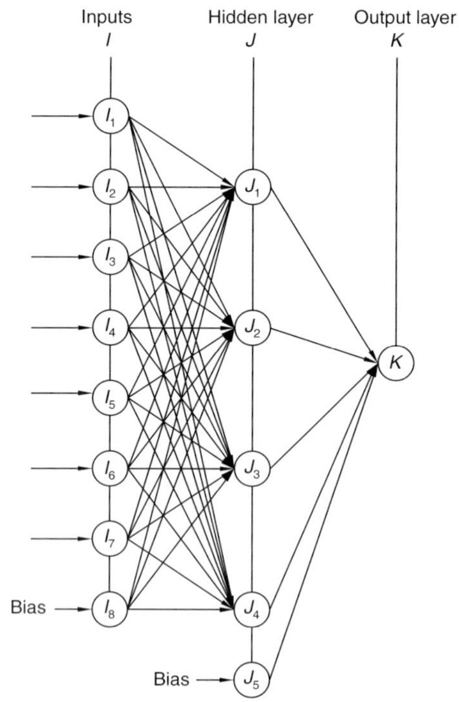


*Fig. 9.9. The neural network for the beam problem in Example 9.1*

*Table 9.3. The weights connecting the layers in the network in Example 9.1*

| Node | $I_1$ | $I_2$ | $I_3$ | $I_4$ | $I_5$ | $I_6$ | $I_7$ | $I_8$ |
|---|---|---|---|---|---|---|---|---|
| $J_1$ | 0.879853 | 2.377706 | –0.46388 | 3.077148 | –0.3432 | –2.30545 | –1.24682 | 0.187463 |
| $J_2$ | 0.317261 | –1.06446 | –2.55922 | –2.80861 | –0.41932 | –2.41553 | –1.78358 | 0.828609 |
| $J_3$ | –0.11594 | –1.84476 | 2.323303 | –2.34725 | 0.122784 | –0.33026 | –0.26004 | 2.260207 |
| $J_4$ | –2.33354 | 1.0814 | 1.99393 | –0.50657 | 2.30448 | –2.24686 | 0.046501 | –0.72225 |

| Node | $J_1$ | $J_2$ | $J_3$ | $J_4$ | $J_5$ |
|---|---|---|---|---|---|
| $K$ | 1.751664 | –4.63334 | –0.02956 | 0.857419 | 0.116395 |

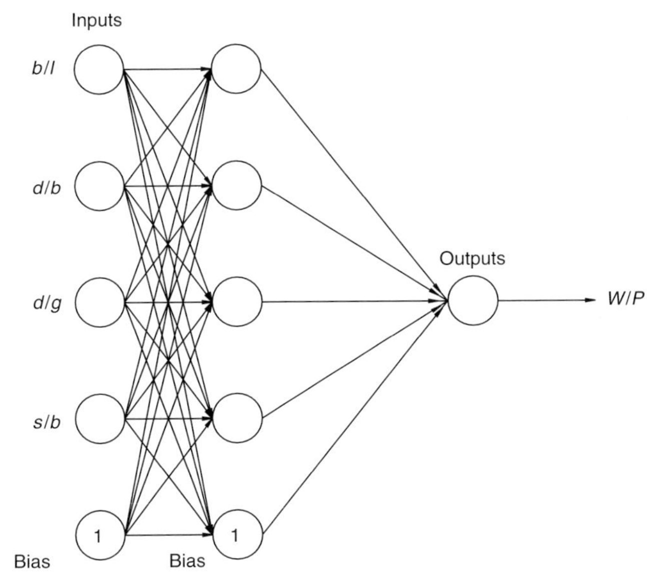


*Fig. 9.10. The neural network for the space frame problem in Example 9.2*

### *Example 9.2*

In this example the neural network is trained by the input and output data available from our archives for 12 space trusses designed in the past with a topology of diagonal top chords over square bottom chords. Using the neural network so trained, the weight of the space truss with the following input data is forecast:

Topology: diagonal over square
Aspect ratio, $b/l = 0.93$
Depth/shorter span, $d/b = 0.05$
Total load acting, $P = 2583.28$ kN
Depth/grid dimension, $d/g = 0.71$
Average column spacing/span, $s/b = 0.50$

A feed-forward error back-propagation network with one hidden layer and sigmoidal squashing function is used (Fig. 9.10).

The neural network is trained by using the available input and output data for 12 space trusses designed in the past. Referring to Table 9.4, the space truss for which we desire the weight appears as item 13 in the table and is easily read as 183.74 kN. To estimate the error, the same space truss was analysed using the SAP90 software and optimized to minimum weight using MEMCO. The optimized weight so found is 193.32 t and the error of 4.95% involved in the neural network prediction is acceptable. This formulation may easily be extended to forecast maximum tension or compression and maximum deflection by adding two more dimensionless outputs $\delta/b$ and $F/W$, where $\delta$ is the maximum deflection and $F$ is the maximum force.

## 9.3. Optimization to minimum weight

### 9.3.1. Optimization by genetic algorithms

Optimization of space trusses to minimum weight is imperative for success in competitive bidding. Recent developments in genetic algorithms [9.10, 9.11] offer a powerful means of optimizing the topology, configuration and member sizes of space trusses. In size optimization, the cross-sectional areas of members are normally chosen as the design variables. The objective function, which is the weight, is to be minimized under certain behavioural constraints on stresses and displacements. The dimensions of the members, such as tubes, that comprise a space truss vary in steps, and hence these dimensions have to be regarded as discrete variables of a discrete set. The resulting discrete optimization problem may be stated as follows:

*Table 9.4. Prediction by neural network of the weights of the space truss in Example 9.2*

| Serial No. | $b/l$ | $d/l$ | $d/g$ | $s/b$ | $W/P$ | $W/P$ predicted by the neural network | Actual weight (kN) | Weight predicted by the neural network (kN) | Difference (%) |
|---|---|---|---|---|---|---|---|---|---|
| 1 | 0.25 | 0.08 | 0.71 | 0.33 | 0.0733447 | 0.0722950 | 332.65 | 327.89 | 1.43 |
| 2 | 0.30 | 0.05 | 0.71 | 0.22 | 0.1078380 | 0.0745675 | 938.72 | 649.79 | 30.85 |
| 3 | 0.50 | 0.09 | 0.71 | 0.40 | 0.0640051 | 0.0700902 | 121.69 | 133.26 | –9.51 |
| 4 | 0.56 | 0.07 | 0.71 | 0.21 | 0.0708151 | 0.0749509 | 182.38 | 193.03 | –5.84 |
| 5 | 0.65 | 0.06 | 0.71 | 0.31 | 0.0730079 | 0.0730207 | 194.04 | 194.07 | –0.02 |
| 6 | 0.67 | 0.13 | 0.50 | 0.54 | 0.0523526 | 0.0633488 | 24.01 | 29.05 | –21.00 |
| 7 | 0.71 | 0.07 | 0.71 | 0.17 | 0.0691972 | 0.0771314 | 141.32 | 157.53 | –11.47 |
| 8 | 0.77 | 0.06 | 0.61 | 0.24 | 0.0679278 | 0.0738129 | 129.68 | 140.91 | –8.66 |
| 9 | 0.82 | 0.07 | 0.61 | 0.27 | 0.0627640 | 0.0734422 | 93.99 | 109.98 | –17.01 |
| 10 | 0.88 | 0.07 | 0.50 | 0.37 | 0.0566839 | 0.0694368 | 51.77 | 63.42 | –22.50 |
| 11 | 0.93 | 0.05 | 0.71 | 0.20 | 0.0773048 | 0.0794790 | 227.58 | 233.98 | –2.81 |
| 12 | 1.00 | 0.08 | 0.50 | 0.50 | 0.0536537 | 0.0677256 | 33.86 | 42.74 | –26.23 |
| 13 | 0.93 | 0.05 | 0.71 | 0.50 | 0.0748337 | 0.0711261 | 193.32 | 183.74 | 4.95 |

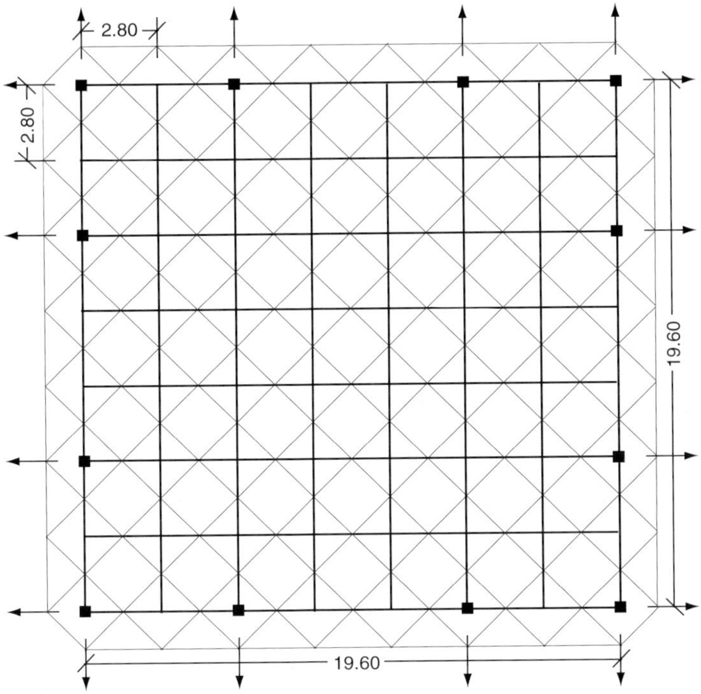


*Fig. 9.11. The double-layer space truss considered in Example 9.3*

$$\begin{aligned}&\text{minimize } F(s)\\&\text{subject to } g_j(s) \le 0, j = 1 \text{ to } m\\&\text{with } s_i \in R^d i = 1 \text{ to } n\end{aligned} \tag{9.7}$$

where $F(s)$ is the objective function and $g_j(s)$ are constraints. $R^d$ is the given set of discrete values that the design variable $s_i$ $(i = 1, 2, 3, \ldots, n)$ can take only from this set.

### 9.3.2. *Elements of genetic algorithms*

Genetic algorithms are inspired by the Darwinian theory of natural selection, which postulates the survival of the fittest. The design variables are represented by binary strings or chromosomes. The optimization process involves the following steps:

- reproduction
- cross-over
- mutation.

For an excellent introduction to these concepts, the reader is referred to Chapter 1 of Goldberg [9.1]. The objective functions resulting from solutions that violate the constraints are penalized by the introduction of a penalty function. Goldberg [9.1] is recommended parallel reading for understanding the following illustrative example.

*Example 9.3*
In this illustrative example the space truss shown in Fig. 9.11 is optimized to minimum weight.

*Geometry*
The topology adopted is diagonally laid top chords over square bottom chords. The space truss is loaded on the top nodes and supported through the bottom nodes to rest on columns. The structural depth is 1.4 m.

*Boundary conditions*
The directions in which the space frame is free to move are indicated by arrows in Fig. 9.11.

*Loading*
Two loading cases are considered:

1. a downward load of 6.68 kN is applied at each of the top chord nodes
2. an upward load of 5.84 kN is applied at each of the top chord nodes.

*Stress analysis*
The stress analysis is elastic; but the loads applied correspond to the ultimate limit state.

*Code of practice*
The design of the members is in accordance with the Indian Standard Code of Practice for General Construction in Steel, IS 800: 1984 (Second Revision). Tubes are assumed to reach their yield strength of 240 MPa in tension at the ultimate limit state. The allowable compressive stress at the ultimate limit state is computed by using the following formula given in IS 800:

$$f_{oa} = (f_{cc} f_y)/[(f_{cc})^n + (f_y)^n]^{1/n} \tag{9.8}$$

where $f_{ao}$ is the permissible compression at the ultimate limit state (in MPa), $f_y$ is the yield stress (in MPa), $f_{cc}$ is the elastic critical stress in compression ($= \pi^2 E/\lambda^2$; $\lambda$ is the ratio of the effective length to the appropriate radius of gyration ($= l/r$)) and $n$ is a factor that is assumed to take the value 1.40. (*Note*: In pin-jointed space trusses, the effective length is taken as 0.85 times the actual length.)

The size optimization problem of a space truss may be formulated as

$$\text{minimize } Z = \sum_{i=1}^{M} l_i \rho_i A_i$$

Subject to the following constraints:

$$\frac{\sigma_{i\,\max}}{\sigma_{\text{allowable}}} - 1 \le 0 \forall_i \in [1, M]$$

and

$$\frac{d_{j\,\max}}{d_{\text{allowable}}} - 1 \le 0 \forall_j \in [1, N]$$

In this case, the deflection constraint imposed was to limit it to a maximum of 60 mm. The following notation is used:

$M$ is the number of members
$l_i$ is the length of member $i$
$N$ is the number of nodes
$\rho_i$ is the density of member $i$
$A_i$ is the cross-sectional area of member $i$
$\sigma_{i\,\max}$ is the maximum stress in member $i$ for all the load cases considered
$\sigma_{i\,\text{allowable}}$ is the allowable stress in member $i$ (tensile or compressive)
$d_{j\,\max}$ is the maximum displacement at node $j$ for all the load cases considered
$d_{j\,\text{allowable}}$ is the maximum allowable displacement.

The space truss considered in this example has 208 nodes and 652 members. The tubular sections considered range in size from 42.5 mm × 3.25 mm to 355 mm × 8.00 mm. Member groups were not assumed *a priori*. Groups were formed after an initial stress analysis and updated every ten generations.

Because genetic algorithms cannot handle constrained optimization problems, the fitness of unfeasible chromosomes needs to be degraded by using a penalty function. Hence the constrained optimization problem is reformulated as

$$\text{minimize } Z = \left\{ \sum_{i=1}^{M} l_i \rho_i A_i \right\} \left\{ 1 + \sum_{j=1}^{M+N} k_j c_j^2 \right\}$$

where

$$c_j = 0 \qquad \text{or} \qquad c_j = \left\{ \frac{\sigma_{i\,\max}}{\sigma_{i\,\text{allowable}}} - 1 \right\}$$

depending on whether

$$\left\{ \frac{\sigma_{i\,\max}}{\sigma_{i\,\text{allowable}}} - 1 \right\} \text{ is} \le 0 \text{ or} > 0 \qquad \text{for all } M \text{ members}$$

Similarly,

$$c_j = 0 \qquad \text{or} \qquad c_j = \left\{ \frac{d_{j\,\max}}{d_{j\,\text{allowable}}} - 1 \right\}$$

depending on whether

$$\left\{ \frac{d_{j\,\max}}{d_{j\,\text{allowable}}} - 1 \right\} \text{ is} \le 0 \text{ or} > 0 \qquad \text{for all } N \text{ joints}$$

In this instance $k_j$ was assumed to be 5.

A hybrid chromosome consisting of two real-coded parts and one binary-coded part was adopted. The first real-coded part contains indices representing the outer diameter (0, 15) and the second real-coded part has indices for thickness corresponding to each outer diameter (0, 3). The binary-coded part consists of an index to the outer diameter (0, 3) if four sections are allowed. During the random initializing of the chromosome, care is taken to ensure that no two sections have the same outside diameter but different thicknesses. Also, the sections are arranged in ascending order of the outer diameter so that, during cross-over, the sections crossing over have approximately the same size.

The optimized weight of 47.36 kN reached at the end of 100 generations involved only four sections. This optimized weight is a little less than the value of 49.302 kN arrived at using the MEMCO software, which uses successive iterations. The results arrived at by means of GA and by means of iteration procedures are compared in Table 9.5.

The convergence of the weight to the optimized value is depicted in Fig. 9.12. This example was worked out in collaboration with a research team at the Indian Institute of Technology, Chennai, India, headed by the (late) Prof. C. S. Krishnamoorthy.

### 9.3.3. Optimization by evolution strategy methods

The most frequently occurring problems relate to the optimization of space trusses to minimum weight under certain behavioural constraints on stresses and displacements. The formulation of this problem has already been discussed in Section 9.3.1.

Evolution strategies were proposed in the 1970s by Rechenberg [9.12] and Schwefel [9.13] for parameter optimization problems. Evolution strategy methods mimic biological evolution in Nature and have the following three characteristics that set them apart from other optimization algorithms:

- they use random operators of *mutation*, *selection* and *recombination* instead of deterministic operators
- instead of focusing on a single design point, they work simultaneously with a population of design points in the space of variables

*Table 9.5. Comparison of the results obtained using a genetic algorithm and an iteration*

| Iteration method | | Genetic algorithm with member grouping | |
|---|---|---|---|
| Section size (mm) | % of total No. of tubes | Section size (mm) | % of total No. of tubes |
| 42.4 × 3.25 | 60.60 | 42.4 × 3.25 | 71.16 |
| 43.8 × 3.25 | 17.10 | 43.8 × 3.25 | 13.50 |
| 60.8 × 3.65 | 13.60 | 60.8 × 3.65 | 12.88 |
| 76.1 × 3.65 | 8.70 | 76.1 × 3.65 | 2.46 |

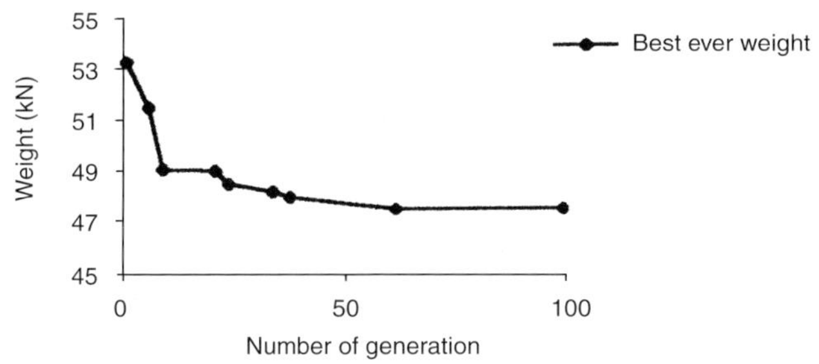


*Fig. 9.12. The convergence to the optimized weight in Example 9.3*

- they can, with some minor modifications, handle continuous, discrete and mixed optimization problems.

The second characteristic renders them suitable for implementation in a parallel computing environment.

Evolution strategy methods have an edge over other optimization algorithms in the following respects:

- When a gradient optimizer is used, the most time-consuming part of the optimization process relates to the sensitivity analysis phase, which is an element of all mathematical programming methods of optimization. Probabilistic search methods such as evolution strategies do not require gradient information, and hence the computationally expensive sensitivity analysis step is obviated [9.14].
- It is widely recognized that probabilistic search methods are generally more robust and present better global behaviour than do mathematical programming methods. They may, however, exhibit a slower rate of convergence towards a global optimum.

### 9.3.4. Evolution strategies for discrete optimization problems

In applying evolution strategy techniques of discrete optimization to space trusses, grouping of members with the same design variables will often be necessary. Such grouping involves a trade-off between the use of more material and the need to meet the requirements of symmetry and uniformity.

For the solution of discrete optimization problems, a modified evolution strategy algorithm has been proposed by Thierauf and Cai [9.15]. The basic differences between discrete and continuous evolution strategies relate to the mutation and recombination operators. The mutation operator ensures that each parent $s_p^{(g)}$ of the current generation produces an offspring $s_o^{(g)}$ the genotype of which is slightly different from that of the parent, so that it is possible to write

$$s_o^{(g)} = s_p^{(g)} + z^{(g)}$$

where $z^{(g)} + (z_1^{(g)}, z_2^{(g)}, \ldots, z_n^{(g)})^{\mathrm{T}}$ is a random vector. The mutation operator in the continuos version of the evolution strategy produces a normally distributed random vector $z^{(g)}$. Each component of this vector has a small standard deviation $\sigma_i$ and a mean value of zero. As a result, it is possible to change each component of the present vector; but usually the changes are small. In the discrete version of the evolution strategy, on the other hand, the random vector $z^{(g)}$ is properly generated to force the offspring vector to move to another set of discrete values. The fact that the difference between two adjacent values can be relatively large is in conflict with the requirement that the variance $\sigma_i^2$ should be small. For this reason, it is suggested that not all the components of a parent vector are randomly changed in every generation; instead, only a few of them, say $l$, are so changed implying that $(n - l)$ components of the random vector $z^{(g)}$ will have zero components. In other words, the terms of the vector $z^{(g)}$ are found as [9.16]

$$z_i^{(g)} = \begin{cases} \kappa\delta_{S_i} & \text{for } l \text{ randomly chosen components} \\ 0 & \text{for } (n-\lambda) \text{ remaining components} \end{cases} \tag{9.9}$$

where $\delta_{S_i}$ is the current difference between two adjacent values in the discrete set, and $\kappa$ is an integer random number with the following distribution:

$$p(\kappa) = \frac{\gamma^k}{\kappa!} e^{-\gamma} \tag{9.10}$$

where $\gamma$ is the standard deviation as well as the expectation of the random number $\kappa$.

For a very small value of $\gamma$, say 0.001, the number $\kappa$ takes the value of zero with a probability greater than 99%, and for $\gamma = 0.05$ the number $\kappa$ takes the value of zero with a probability of 95% and the value of 1.0 with a probability of 5%. Thus the random change $z_i^{(g)}$ is controlled by the parameter $\gamma$. A uniformly distributed random choice decides which $l$ components are to be changed according to equation (9.9). The choice of $l$ depends on the size of the problem and it is usually taken as one-fifth of the total number of design variables.

In both continuous as well as discrete versions of the evolution strategy, there are two different types of selection:

- the $\mu + \lambda$ scheme, where the best $\mu$ individuals are selected from a temporary population of $\mu + \lambda$ individuals of parents and offspring to form the parents of the next generation
- the $(\mu, \lambda)$ scheme, where the $\mu$ individuals produce $\lambda$ offspring $(\mu \le \lambda)$ and the selection process defines a new population of $\mu$ individuals from the set of $\lambda$ offspring only.

In the second case, the life of each individual is limited to one generation. The $(\mu, \lambda)$ scheme is reported to perform better where the objective function is noisy [9.17].

The evolution strategy optimization procedure starts with a set of parent vectors, and if any of these parent vectors leads to an unfeasible design it is modified until it becomes feasible. The offspring generated are also checked to ensure that they are within the feasible region. According to the $\mu + \lambda$ scheme, in every generation the values of the objective function of parent and offspring vectors are compared, the worst vectors are dropped and those which survive become the parents of the next generation. On the other hand, in the $(\mu, \lambda)$ scheme the offspring vectors of each generation become the parents of the next generation. This procedure is repeated until the termination criterion (see Section 9.3.4.1) is satisfied. The computational efficiency of the evolution strategy optimization procedure is affected by the number of parents and offspring involved. It has been observed that the best results are achieved by choosing the number of parents and offspring to be the same as the number of design variables [9.18].

### *9.3.4.1. Termination criteria*

For discrete optimization problems the procedure is terminated when one of the following heuristic criteria is satisfied [9.19]:

- when the best value of the objective function remains unchanged in the last $4n\mu/\lambda$ generations
- when the mean value of the objective function from all parent vectors in the last $2n\mu/\lambda$ generations has not improved by less than a given value, say 0.0001
- when the relative difference between the best objective function value and the mean of the values of all the objective functions of the parent vectors in the current generation is less than a specified value, say 0.0001
- when the ratio $\mu_o/\mu$ has reached a given value between 0.5 and 0.80, where $\mu_o$ is the number of parent vectors in the current generation with the best objective function value.

### *9.3.4.2. Outline of the evolution strategy optimization procedure*

1. Selection step: select $s_i$ $(i = 1, 2, \ldots, \mu)$ parent vectors of the design variables.
2. Analysis step: solve $K(s_i)u_i = f$ $(i = 1, 2, \ldots, \mu)$.
3. Constraints check: check to ensure that all the parent vectors are in the feasible region.
4. Offspring generation: generate $s_j$ $(j = 1, 2, \ldots, \lambda)$ offspring vectors of the design variables.
5. Analysis step: solve $K(s_j)u_j = f$ $(j = 1, 2, \ldots, \lambda)$.
6. Constraints check: if satisfied, continue; otherwise, change $s_j$ and go to step 4.
7. Selection step: select the next generation of parents according to the $\mu + \lambda$ or $(\mu, \lambda)$ scheme.
8. Convergence check: if the termination criteria are satisfied, stop.

This step-by-step procedure is illustrated below by application to the optimization to minimum weight of a three-bar truss. This example has been used in numerous investigations since its first introduction by Schmit [9.20]. The final design is required to be symmetric.

### *Example 9.4*

The aim is to use the evolution strategy procedure to optimize to minimum weight the three-bar pin-jointed truss shown in Fig. 9.13. The design data are as follows:

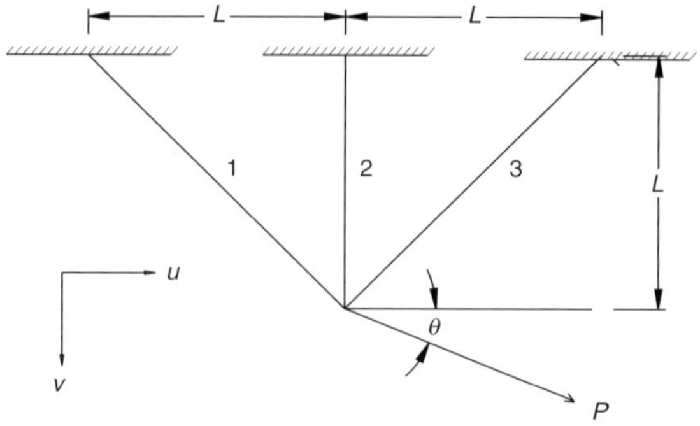


*Fig. 9.13. The three-bar pin-jointed truss considered in Example 9.4*

Allowable stress: members 1 and 3, $\sigma_{1a} = \sigma_{3a} = 5000$ psi; member 2, $\sigma_{2a} = 20\,000$ psi
Allowable displacement: $\Delta_u = 0.005$ in., $\Delta_v = 0.005$ in.
Modulus of elasticity: $E = 10^7$ psi
Weight density: $\rho = 0.1$ lb/in.$^3$
Lower limits of the design variables: (1.0, 1.0, 1.0) in.$^2$
Upper limits of the design variables: (20.0, 20.0, 20.0) in.$^2$
Parameter: $L = 1.0$ in.
Load: $P = 40\,000$ lb
Angle: $\theta = 45°$

*Design variables*
The design variables are $A_1$ and $A_2$, where $A_1$ is the cross-sectional area of bars 1 and 3 and $A_2$ is the cross-sectional area of bar 2.

*Objective function*
The objective is to minimize the weight, which is equivalent to minimizing the volume:

$$\text{volume} = L(2\sqrt{2}A_1 + A_2) \tag{9.11}$$

*Stresses in members*
The resolved components of the force in the $u$ and $v$ directions are

$$P_u = P\cos\theta \tag{9.12a}$$

$$P_v = P\sin\theta \tag{9.12b}$$

By carrying out an analysis of the three-bar truss as a statically indeterminate structure, the stresses in members 1, 2 and 3 are

$$\sigma_1 = \frac{1}{\sqrt{2}}\left[\frac{P_u}{A_1} + \frac{P_v}{A_1 + \sqrt{2}A_2}\right] \tag{9.13}$$

$$\sigma_2 = \frac{\sqrt{2}P_v}{A_1 + \sqrt{2}A_2} \tag{9.14}$$

$$\sigma_3 = \frac{1}{\sqrt{2}}\left[\frac{P_v}{A_1 + \sqrt{2}A_2} - \frac{P_u}{A_1}\right] \tag{9.15}$$

From equations (9.13) and (9.15) it can be seen that $\sigma_1 > \sigma_3$ and hence only $\sigma_1$ needs to be considered when formulating the constraints.

*The constraints on the stresses*

Members may fail in tension, crushing or buckling. Hence the normalized constraints on stresses may be stated as:

- Stress in member/allowable stress < 1, to prevent failure by tension or crushing; the allowable stress is taken as $0.6 f_y$, where $f_y$ is the yield stress.
- Compressive stress in member/buckling stress $\sigma_b < 1$.

For a thin tube of diameter $D$ and thickness $t$, and ratio $D/t$ in the range 10 to 20, it is sufficiently accurate to compute the buckling stress as

$$\sigma_b = \frac{\pi^2 E}{8L^2}(D-t)^2 \tag{9.16}$$

where $L$ is the length.

The constraint on buckling may therefore be stated as

$$|\text{compressive stress}| \leq \sigma_b$$

The absolute value of the compressive stress in a member is less than or equal to the buckling stress.

*The constraint on the deflections*

The deflections in the $u$ and $v$ directions are computed as

$$u = \frac{\sqrt{2}LP_u}{A_1 E} \tag{9.17}$$

$$v = \frac{\sqrt{2}LP_v}{(A_1 + \sqrt{2}A_2)E} \tag{9.18}$$

where $E$ is the modulus of elasticity of the material. The constraint on deflections may be stated as

$$u \leq \Delta u \qquad \text{and} \qquad v \leq \Delta v$$

where $\Delta u$ and $\Delta v$ are the permissible deflections in the $u$ and $v$ directions.

*Solution*

The $\mu + \lambda$ system of selection is used in this example with $\mu = \lambda = 5$. The current difference $\delta s_i$ is taken as 0.20 and $l = 1$. The recombination scheme used involves combining the best member of the parent population with a randomly chosen member from the same population.

The procedure is terminated when the mean value of the objective functions from all parent vectors has not improved by less than a specified value, say 0.001 in the last four generations. To illustrate the procedure, the evolution strategy process will be carried through three generations.

Referring to Table 9.6:

- Recombine member 5 with member 1 to obtain the temporary offspring (8, 10, 8). Mutate the design variable $A_1$ with $\kappa = 0$ in equation (9.9) to produce the first feasible offspring (8, 10, 8) with an objective function value of 32.6 lb.
- Recombine member 5 with member 5 to obtain the second temporary offspring (8, 7, 8). Mutate the design variable $A_2$ with $\kappa = 1$ to produce the corresponding feasible offspring (8, 7.2, 8) with an objective function value of 29.8 lb.
- Recombine member 5 with member 4 to obtain the third temporary offspring vector (8, 12, 8). Mutate the design variable $A_1$ with $\kappa = 1$ to produce the corresponding feasible offspring (7.8, 12, 7.8) with an objective function value of 34.1 lb.
- Recombine member 5 with member 3 to obtain the fourth temporary offspring (14, 7, 4). Mutate design variable $A_1$ with $\kappa = 1$ to produce the corresponding feasible offspring (13.8, 7, 13.8) with an objective function value of 46.0 lb.
- Recombine member 5 with member 1 to obtain the fifth temporary offspring (10, 7, 10). Mutate design variable $A_2$ with $\kappa = 1$ to produce the corresponding feasible offspring (10, 6.8, 10) with an objective function value of 35.1 lb.

If an offspring turns out to be unfeasible, the best member is recombined with a randomly chosen member and the mutation step carried out to generate a feasible offspring. The second-generation population is given in Table 9.7.

*Table 9.6. The first generation of parent vectors in Example 9.4*

| Member No. | Design point (in.$^2$) | Objective function |
|---|---|---|
| 1 | (10, 10, 10) | 38.3 |
| 2 | (14, 13, 14) | 52.6 |
| 3 | (14, 14, 14) | 53.6 |
| 4 | (8, 12, 8) | 34.6 |
| 5 | (8, 7, 8) | 29.6 |

*Table 9.7. The second generation of parent vectors in Example 9.4*

| Member No. | Design point (in.$^2$) | Objective function |
|---|---|---|
| 1 | (8, 7, 8) | 29.6 |
| 2 | (8, 12, 8) | 34.6 |
| 3 | (8, 10, 8) | 32.6 |
| 4 | (8, 7.2, 8) | 29.8 |
| 5 | (7.8, 12, 7.8) | 34.1 |

*Table 9.8. The third generation of parent vectors in Example 9.4*

| Member No. | Design point (in.$^2$) | Objective function |
|---|---|---|
| 1 | (7.8, 7, 7.8) | 29.1 |
| 2 | (8, 7, 8) | 29.6 |
| 3 | (8, 7.2, 8) | 29.8 |
| 4 | (7.8, 7, 7.8) | 29.1 |
| 5 | (8, 7, 8) | 29.6 |

We now commence the next cycle:

- Recombine member 1 with member 5 to obtain the temporary offspring (7.8, 7, 7.8). Mutate design variable $A_1$ with $\kappa = 0$ to produce the corresponding feasible offspring (7.8, 7, 7.8) with an objective function value of 29.1 lb.
- Recombine member 1 with member 2 to obtain the temporary offspring (8, 7, 8). Mutate design variable $A_2$ with $\kappa = 0$ to produce the feasible offspring (8, 7, 8) with an objective function value of 29.6 lb.
- Recombine member 1 with member 4 to obtain the temporary offspring (8, 7.2, 8). Mutate design variable $A_2$ with $\kappa = 0$ to produce the feasible offspring (8, 7.2, 8) with an objective function value of 29.8 lb.
- Recombine member 1 with member 1 to obtain the temporary offspring (8, 7, 8). Mutate design variable $A_1$ with $\kappa = 1$ to generate the feasible offspring (7.8, 7, 7.8) with an objective function value of 29.1 lb.
- Recombine member 1 with member 3 to obtain temporary offspring (8, 7, 8). Mutate design variable $A_1$ with $\kappa = 0$ to produce the feasible offspring (8, 7, 8) with an objective function value of 29.6 lb.

The third-generation population is given in Table 9.8.

The procedure is repeated until the termination criterion is satisfied. In the current example the optimized design of (5.8, 5.2, 5.8) with an objective function value of 21.6 lb is achieved in 14 generations involving 86 finite element analyses.

### 9.3.5. *Optimization by the neural network–evolution strategy hybrid*

This is a hybrid method that combines the neural network and evolution strategy techniques with the aim of improving robustness and computational efficiency [9.21]. For each optimization problem, the neural network is first trained using information generated from a number of carefully selected analyses. The data from these analyses are processed to provide input and output pairs, which are needed to train the network. The trained neural network is then used to predict the response of the structure in terms of the objective function and constraint function values corresponding to different design variables. The neural network architecture used is the standard feed-forward error back-propagation training algorithm.

The merits of the method are best demonstrated by its application to the problem of optimizing a large space truss. The example chosen is the space truss of the long span aircraft hangar recently built at Mumbai for Air India [9.22], for which the first

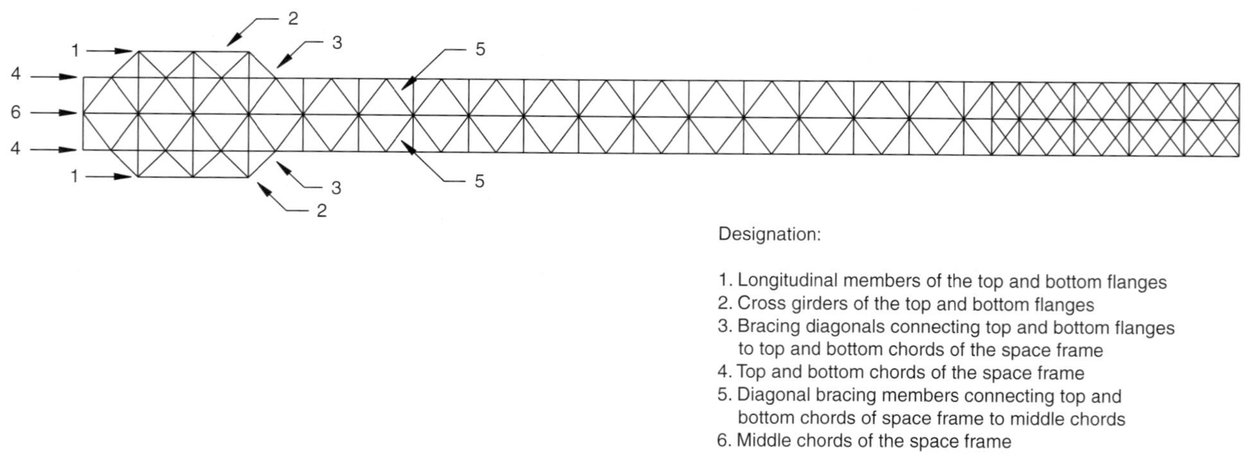


*Fig. 9.14. Section of the Air India hangar considered in Example 9.5*

author developed the conceptual design. The weight and the maximum deflections for this preliminary design were made available by the first author to Professor Manolis Papadrakakis and Dr Nikolaos D. Lagaros at the National University at Athens who carried out the evolution strategy and the neural network–evolution strategy (NNES) optimization.

*Example 9.5*

Our aim is to optimize the space truss of the Air India hangar at Mumbai to minimum weight.

*Description of the hangar*

The hangar is a triple-layer space frame with a five-layer front girder spanning 130.884 m. The front girder is in fact an extension of the space frame, formed by adding two layers, one at the top and the other at the bottom (Fig. 9.14). These heavily stressed top and bottom layers consisting of longitudinal and cross girders, formed of closed box sections, will be here designated as 'flanges'. The diagonal members connecting the top and bottom flanges to the top and bottom chords of the space frame are also fabricated box sections. The hangar was considered an ideal candidate to test the robustness and computational efficiency claimed for the NNES optimization technique.

The members of the space truss were grouped as follows:

- Group 1: the longitudinal members of the top and bottom flanges.
- Group 2: the cross girders of the top and bottom flanges.
- Group 3: the bracing diagonals connecting the top and bottom flanges to the top and bottom chords of the space truss.
- Group 4: the top and bottom chords of the space truss.
- Group 5: the diagonal bracing members connecting the top and bottom chords of the space frame to the middle chords.
- Group 6: the middle chords of the space truss.

Taking advantage of symmetry, only one half of the hangar, comprising 1807 nodes and 6487 members, was considered. Members of groups 1 to 3 were selected from the box sections listed in Table 9.9. Members of groups 4 to 6 were selected from the tubes listed in Table 9.10. In addition to the constraints on the stresses, a constraint of 750 mm maximum deflection was imposed. Optimization was carried out using both the evolution strategy and the NNES technique.

In the evolution strategy procedure, the best of five parents and five offspring was used (Table 9.11). Both techniques led to an optimum weight of 431.39 t for one half of the hangar and a maximum deflection of 637 mm. These values are in good agreement with the weight of 439.43 t and the maximum deflection of 624 mm computed by the first author in his preliminary design. The optimized design obtained using the evolution strategy and the NNES procedure is represented by the design variables [17(s), 7(s), 7(s), 4(t), 5(t), 1(t)], where members 17(s) and 7(s) correspond to structural box sections selected from Table 9.9 with serial numbers 17

*Table 9.9. Example 9.5: closed box structural sections*

| Member No. | Designation | Description | Area ($mm^2$) |
|---|---|---|---|
| 1 | ISMC 100 | Single channel | 1 170 |
| 2 | 2 × ISMC 75 | Closed box section made up of two channels | 1 740 |
| 3 | 2 × ISMC 100 | Closed box section made up of two channels | 2 340 |
| 4 | 2 × ISMC 125 | Closed box section made up of two channels | 3 238 |
| 5 | 2 × ISMC 150 | Closed box section made up of two channels | 4 176 |
| 6 | 2 × ISMC 175 | Closed box section made up of two channels | 4 878 |
| 7 | 2 × ISMC 200 | Closed box section made up of two channels | 5 642 |
| 8 | 2 × ISMC 225 | Closed box section made up of two channels | 6 802 |
| 9 | 2 × ISMC 250 | Closed box section made up of two channels | 7 734 |
| 10 | 2 × ISMC 300 | Closed box section made up of two channels | 9 128 |
| 11 | 2 × ISMC 350 | Closed box section made up of two channels | 10 732 |
| 12 | 2 × ISMC 400 | Closed box section made up of two channels | 12 585 |
| 13 | 2 × ISMC 400 with 2 mm × 8 mm thick MS plates | Closed box section made up of two channels with two plates welded at top and bottom | 14 986 |
| 14 | 2 × ISMC 400 with 2 mm × 12 mm thick MS plates | Closed box section made up of two channels with two plates welded at top and bottom | 16 186 |
| 15 | 2 × ISMC 400 with 2 mm × 16 mm thick MS plates | Closed box section made up of two channels with two plates welded at top and bottom | 17 386 |
| 16 | 2 × ISMC 400 with 2 mm × 25 mm thick MS plates | Closed box section made up of two channels with two plates welded at top and bottom | 20 086 |
| 17 | 4 × ISMC 400 | Closed double box section made up of four channels | 25 172 |
| 18 | 4 × ISMC 400 with 2 mm × 8 mm thick MS plates | Closed double box section made up of four channels with two plates welded at top and bottom | 30 772 |
| 19 | 4 × ISMC 400 with 2 mm × 16 mm thick MS plates | Closed double box section made up of four channels with two plates welded at top and bottom | 36 372 |
| 20 | 4 × ISMC 400 with 2 mm × 20 mm thick MS plates | Closed double box section made up of four channels with two plates welded at top and bottom | 39 172 |
| 21 | 4 × ISMC 400 with 2 mm × 25 mm thick MS plates | Closed double box section made up of four channels with two plates welded at top and bottom | 42 672 |
| 22 | 4 × ISMC 400 with 2 mm × 32 mm thick MS plates | Closed double box section made up of four channels with two plates welded at top and bottom | 47 572 |
| 23 | 4 × ISMC 400 with 4 mm × 20 mm thick MS plates | Closed double box section made up of four channels with four plates welded at top and bottom | 51 172 |
| 24 | 4 × ISMC 400 with 4 mm × 25 mm thick MS plates | Closed double box section made up of four channels with four plates welded at top and bottom | 57 672 |

*Table 9.9. Contd*

| Member No. | Designation | Description | Area ($mm^2$) |
|---|---|---|---|
| 25 | 4 × ISMC 400 with 4 mm × 32 mm thick MS plates | Closed double box section made up of four channels with four plates welded at top and bottom | 66 772 |
| 26 | 4 × ISMC 400 with 4 mm × 40 mm thick MS plates | Closed double box section made up of four channels with four plates welded at top and bottom | 77 172 |
| 27 | 4 × ISMC 400 with 4 mm × 50 mm thick MS plates | Closed double box section made up of four channels with four plates welded at top and bottom | 90 172 |

ISMC, Indian Standard Metric Channels

*Table 9.10. Example 9.5: tubular sections*

| Member No. | Outer diameter, $D$ (mm) | Thickness, $T$ (mm) | Area ($mm^2$) | Weight (kg/m) |
|---|---|---|---|---|
| 1 | 60.30 | 3.25 | 582.73 | 4.57 |
| 2 | 76.10 | 4.50 | 1012.63 | 7.95 |
| 3 | 88.90 | 4.85 | 1281.16 | 10.06 |
| 4 | 114.30 | 5.40 | 1848.19 | 14.51 |
| 5 | 139.70 | 5.40 | 2279.26 | 17.89 |
| 6 | 152.40 | 5.40 | 2494.8 | 19.58 |
| 7 | 165.10 | 5.40 | 2710.34 | 21.28 |
| 8 | 193.70 | 5.90 | 3482.35 | 27.34 |
| 9 | 219.10 | 5.90 | 3953.34 | 31.03 |
| 10 | 273.00 | 5.90 | 4952.8 | 38.88 |

and 7, and 4(t), 5(t) and 1(t) represent tubes with serial numbers 4, 5 and 1 in Table 9.10. The letters 's' and 't' stand for structural box sections and tubes, respectively.

The results obtained using the evolution strategy and NNES are shown in Tables 9.12 and 9.13. It can be clearly seen that the hybrid NNES optimization technique is computationally more efficient.

## 9.4. Expert systems

Expert systems are proposed as a means of transferring the expertise of a few to the many, so that even relatively inexperienced engineers who have no opportunities of directly interacting with experts can produce acceptable designs of complex structures such as space frames.

### *9.4.1. What are expert systems?*

According to Feigenbaum [9.23]: 'An expert system is an interactive computer program incorporating judgement, experience, rules of thumb, intuition and other expertise to provide knowledgeable advice about a variety of tasks'. A more elaborate definition offered by Weiss and Kulikowski [9.24] states: 'an expert system solves real world, complex problems using a complex model of expert human reasoning, reaching the same conclusions that the human expert would reach if faced with a comparable problem'.

Stated briefly, expert systems are intelligent computer programs with a *knowledge base* incorporating the best available knowledge in a specialized *domain*.

A well-conceived expert system can be expected to emulate the decision-making skills of an acknowledged expert practitioner in that field. The knowledge base of an expert system comprises facts and heuristics. *Facts* constitute the large body of easily accessible and widely shared information, which may be found in books, publications and codes of practice. *Heuristics* are certain procedures, which are known to work in practice, but cannot be proved formally. These include rules of thumb based on intuition and judgement that an expert utilizes in decision-making. *Knowledge acquisition* is the first step involved in building an expert system. This task is undertaken by the *knowledge engineer*, who scans the technical literature (e.g. books, manuals and journals), interviews domain experts and reviews available

*Table 9.11. Example 9.5: the sizes of the initial parent vectors used in the optimization by means of the evolution strategy*

| | Design | Weight (t) |
|---|---|---|
| *Test set 1* | | |
| Parent 1 | 24(s), 16(s), 16(s), 10(t), 10(t), 10(t) | 1169.19 |
| Parent 2 | 24(s), 16(s), 16(s), 10(t), 10(t), 10(t) | 1169.19 |
| Parent 3 | 24(s), 16(s), 16(s), 10(t), 10(t), 10(t) | 1169.19 |
| Parent 4 | 24(s), 16(s), 16(s), 10(t), 10(t), 10(t) | 1169.19 |
| Parent 5 | 24(s), 16(s), 16(s), 10(t), 10(t), 10(t) | 1169.19 |
| *Test set 2* | | |
| Parent 1 | 20(s), 13(s), 12(s), 6(t), 5(t), 5(t) | 631.35 |
| Parent 2 | 21(s), 8(s), 15(s), 9(t), 5(t), 10(t) | 815.03 |
| Parent 3 | 24(s), 14(s), 14(s), 7(t), 5(t), 5(t) | 735.29 |
| Parent 4 | 22(s), 15(s), 10(s), 5(t), 6(t), 6(t) | 639.55 |
| Parent 5 | 21(s), 13(s), 15(s), 4(t), 9(t), 7(t) | 774.42 |
| *Test set 3* | | |
| Parent 1 | 20(s), 13(s), 12(s), 7(t), 5(t), 9(t) | 681.25 |
| Parent 2 | 21(s), 12(s), 15(s), 9(t), 5(t), 10(t) | 827.30 |
| Parent 3 | 24(s), 13(s), 14(s), 7(t), 5(t), 6(t) | 737.45 |
| Parent 4 | 22(s), 14(s), 15(s), 5(t), 6(t), 6(t) | 708.09 |
| Parent 5 | 21(s), 13(s), 9(s), 4(t), 9(t), 7(t) | 691.32 |
| *Test set 4* | | |
| Parent 1 | 23(s), 13(s), 13(s), 6(t), 5(t), 5(t) | 688.99 |
| Parent 2 | 24(s), 13(s), 15(s), 9(t), 5(t), 10(t) | 878.63 |
| Parent 3 | 24(s), 14(s), 14(s), 7(t), 5(t), 5(t) | 735.29 |
| Parent 4 | 22(s), 15(s), 10(s), 5(t), 6(t), 6(t) | 639.55 |
| Parent 5 | 21(s), 13(s), 15(s), 4(t), 9(t), 7(t) | 774.42 |

*Table 9.12. Example 9.5: results of the optimizations by means of the evolution strategy and the finite element (FE) analysis*

| Test | Optimum weight (t) | Optimum design | No. of generations | No. of FE analyses | Maximum deflection (mm) | Time (s) |
|---|---|---|---|---|---|---|
| 1 | 431.39 | 17s, 7s, 7s, 4t, 5t, 1t | 65 | 213 | 637 | 10 329 |
| 2 | 431.39 | 17s, 7s, 7s, 4t, 5t, 1t | 32 | 100 | 637 | 4 875 |
| 3 | 431.39 | 17s, 7s, 7s, 4t, 5t, 1t | 41 | 123 | 637 | 5 989 |
| 4 | 431.39 | 17s, 7s, 7s, 4t, 5t, 1t | 56 | 159 | 637 | 7 727 |

*Table 9.13. Example 9.5: results of the optimization by means of the NNES procedure*

| Test | Optimum weight (t) | Optimum design | No. of generations | No. of FE analyses for the neural network training set | No. of FE analyses using the neural network | Maximum deflection (mm) | Time (s) |
|---|---|---|---|---|---|---|---|
| 1 | 431.39 | 17s, 7s, 7s, 4t, 5t, 1t | 65 | 60 | 221 | 634 | 2956.21 |
| 2 | 431.39 | 17s, 7s, 7s, 4t, 5t, 1t | 32 | 60 | 99 | 632 | 2930.19 |
| 3 | 431.39 | 17s, 7s, 7s, 4t, 5t, 1t | 41 | 60 | 127 | 640 | 2936.16 |
| 4 | 431.39 | 17s, 7s, 7s, 4t, 5t, 1t | 56 | 60 | 151 | 639 | 2941.28 |

experimental data. The knowledge so acquired has to be represented in a usable form. The most favoured approach is to use a series of production rules. Each rule consists of an ordered pair of symbols (IF and THEN). The first constitutes the premises or conditions and the second represents actions.

### *9.4.2. Evolution of expert systems*

Expert systems are part of the large body of knowledge referred to as 'artificial intelligence'. Expert systems, as we know them today, have their origins in the pioneering work of Feigenbaum, Lederberg, Shortliffe and Buchanan at Stanford University in the late 1960s and early 1970s. The earliest systems to appear were MYCIN for diagnosing infectious diseases, PROSPECTOR for interpreting geological information and MOLGEN for planning molecular genetic experiments.

### *9.4.3. The component parts of an expert system*

The three essential parts of an expert system are a knowledge base, an inference engine and a user interface. The inference engine represents the domain-independent part of the system, and consists of a logical and a computational module. It receives and processes information from the knowledge base, rule base, database and the user interface to arrive at decisions.

It is convenient to divide the frame of an expert system into a domain-dependent knowledge base and a non-domain-dependent inference engine [9.25].

The non-domain-dependent part, representing the tools used to build an expert system, is known as the *shell*. Shells lend themselves to repeated use in different domains of knowledge. The shell used in building the CCC Xpert System described in this chapter is developed on Visual Rule Studio (VRS) [9.26] using Production Rule Language (PRL).

### 9.4.4. Visual Rule Studio

The VRS is not just an upgrade of LEVEL 5 OBJECT, but represents a generational leap in the application of expert system technology to the problem of rules representation. First-generation shells such as LEVEL 5 OBJECT cannot incorporate currently available emerging technologies such as OLE and COM. VRS is developed to form an integral part of Microsoft Visual Basic 5.0/6.0, which is today's proven component-oriented development tool. This enables the developer fully to leverage OLE and COM technologies. The developer can design the graphical user interface (GUI) to suit his or her specific requirements.

VRS supports three types of inferencing strategy:

- backward chaining
- forward chaining
- hybrid chaining.

The inference engine automatically searches and selects from the available rules to process the information.

### 9.4.5. Features of the CCC Xpert System

- The multiple-document interface (MDI) is incorporated in the CCC Xpert System. This permits the user to create multiple sessions and compare various sessions at the same time, if required.
- The user has the option of saving the worked sessions for later retrieval. The sessions can be saved in *.sfm format.
- The CCC Xpert System has the same look and feel as standard Microsoft applications and has features such as window tiling, tools and status bar, print set-up and print preview.
- The CCC Xpert System is mouse-enabled and several hot keys and shortcut keys are provided for easy navigation.
- Detailed help with several tutorials is available on line.
- The system is a stand-alone executable project that can be distributed easily.

### 9.4.6. The design process

The design of a flat space frame involves three distinct phases:

- preliminary design
- structural analysis of the space frame to determine the forces, displacements and reactions
- design of the members and optimization of the space frame to minimum weight.

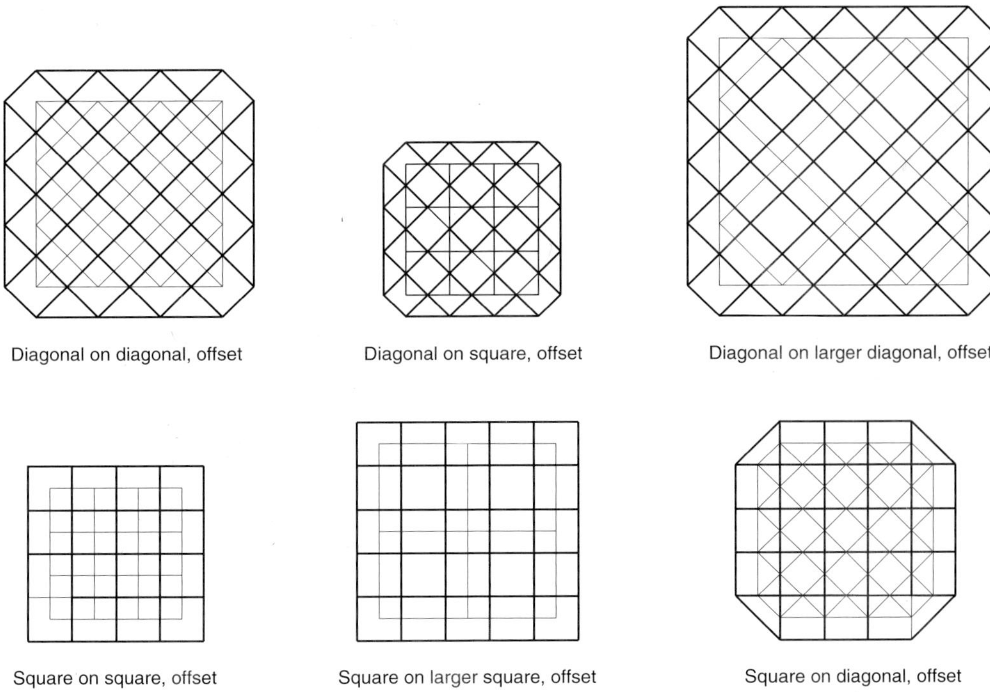


*Fig. 9.15. The six topologies commonly used for space frames*

### *9.4.6.1. Preliminary design*

In a statically determinate structure it is not necessary to know the member sizes in order to start structural analysis. However, in an indeterminate structure, such as a space frame, preliminary member sizes have to be chosen to begin the first cycle of analysis. This is often done by replacing the space frame by the equivalent plate or grid, for which formulae for forces and deflections are readily available. When the forces in the members have been so estimated, the appropriate sizes to carry them safely can be readily found. The other design decisions involved in the preliminary phase are:

- choice of purlin spacing
- choice of grid
- choice of topology
- choice of structural depth.

The roof cladding specified determines the purlin spacing, which in turns governs the choice of the grid and structural depth. The six topologies that are commonly used (Fig. 9.15) are:

- diagonal on diagonal, offset
- diagonal on larger diagonal, offset
- diagonal on square, offset

- square on diagonal, offset
- square on larger square, offset
- square on square, offset.

#### 9.4.6.2. *Structural analysis*

The first cycle of the structural analysis is carried out by using the preliminary sizes for the members, arrived at by means of the plate or grid analogy, to determine the member forces and displacements.

#### 9.4.6.3. *Optimization to minimum weight*

The program MEMCO is used to optimize the space frame to minimum weight. The program scans the forces in the members found in the first cycle of analysis and chooses the appropriate sizes of members to match the forces. It automatically rewrites the input data in SAP format and carries out another cycle of analysis. At the end of each cycle the weight of the space frame is computed and displayed. The iteration is continued until the difference in weight between two successive cycles is negligible or when the designer decides to stop it.

### 9.4.7. *The scope and capabilities of the CCC Xpert System*

The expert system has the capability to carry out preliminary design, structural design of members and optimization of the weight of the structure to the minimum. The CCC Xpert System comprises the VRS, SAP90 stress analysis program and the optimization program entitled MEMCO.

The knowledge base of the CCC Xpert System comprises:

- commercially available circular hollow sections and rectangular hollow sections and their properties (outer diameter, thickness and area)
- the permissible tension and compression in members at the ultimate limit state computed according to the Eurocode (No. 3: 1989), the British Standard (BS 5950: 1990) and the Indian Code (IS 800: 1984)
- formulae given in the Indian Code (IS 800: 1984) for permissible deflections
- formulae for the maximum forces in equivalent plates or grids corresponding to the selected topology
- the Xpert system is written in the limit state format, and so the loads corresponding to the ultimate limit state are applied at the nodes, but the stress analysis is elastic
- given the length, breadth, the prescribed roof cladding, loading and support conditions, the VRS displays
  - the selected topology
  - the selected structural depth
  - the estimated maximum force and deflection
  - the size of the top and bottom chords to be assumed for the first cycle of analysis.

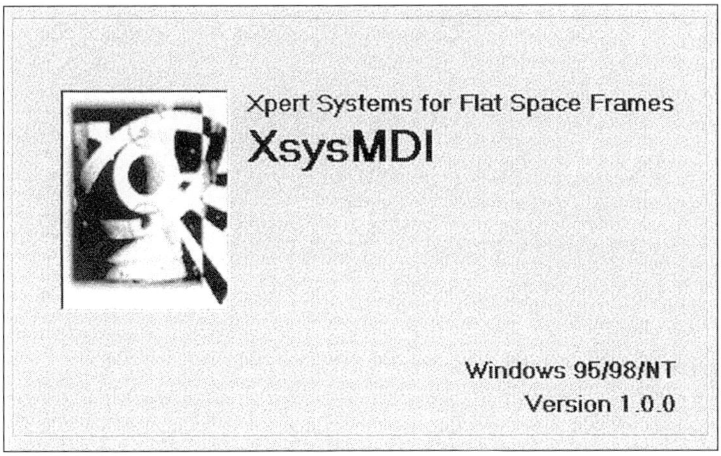


*Fig. 9.16. The CCC Xpert System flash screen*

The structural analysis and optimization are carried out externally by means of SAP90 and MEMCO.

## 9.5. Illustrative example

The tutorial for the CCC Xpert System demonstrates a step-by-step procedure for operating the system, with the aid of a worked example. There are several pages (Input, Space Frame Data, Design Criteria, Pre-Design, and Summary). When the user starts the application, a flash screen appears (Fig. 9.16).

Starting with the **Input** page, carry out the following steps:

1. Select **Acrylic Dome** as the roof cladding from the list available in the drop-down menu **Cladding Type**. A purlin spacing of 1.5 m is appropriate for the selected roof cladding.
2. Under the **Loading Type Frame**, select moderate loading by clicking on the **Moderate** option button (Fig. 9.17).
3. Select the close support condition from the **Support Condition** frame by clicking on the **Close** option button. (Fig. 9.17).
4. Now move to the next page **Space Frame Data**. The CCC Xpert System recommends square on square, offset topology on the basis of the conditions entered in the Input page. However, the user can override this choice and select any other available topology (Fig. 9.18).
5. Enter the length and breadth of the space frame (in meters) in the **Size** frame as 27 m and 24 m, respectively. There is a built-in error control, which prevents the user from entering a breadth that is higher than the length (Fig. 9.18).
6. In the **Data** frame, the grid size of 1.5 m is automatically arrived at from the selected cladding type, and the structural depth of 1.0607 m is calculated from the grid size. The user can override these values by entering a different grid size. The structural depth is dynamically updated. Similarly, the structural depth can also be changed (Fig. 9.18).

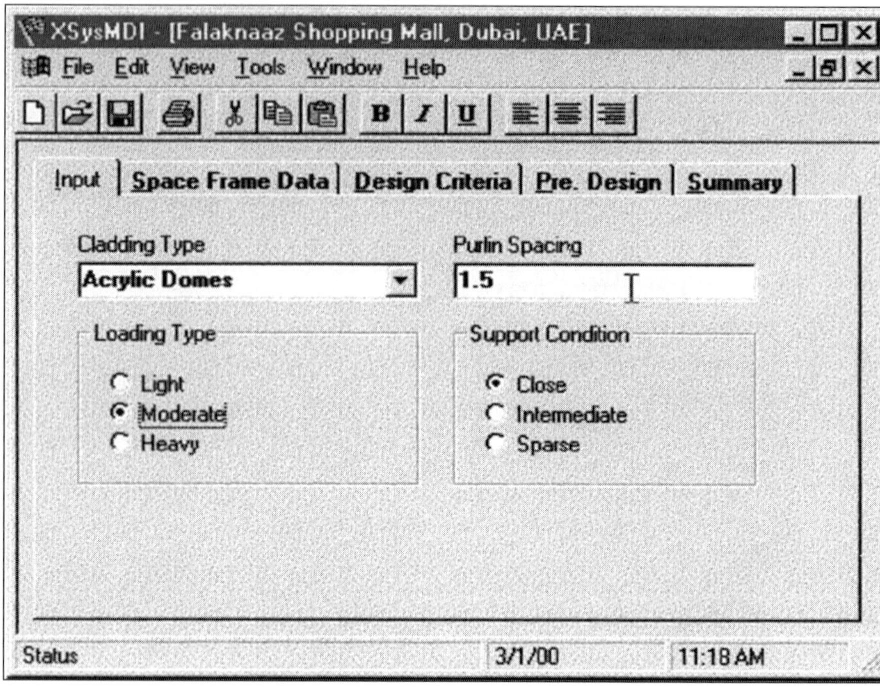


*Fig. 9.17. The CCC Xpert System Input page*

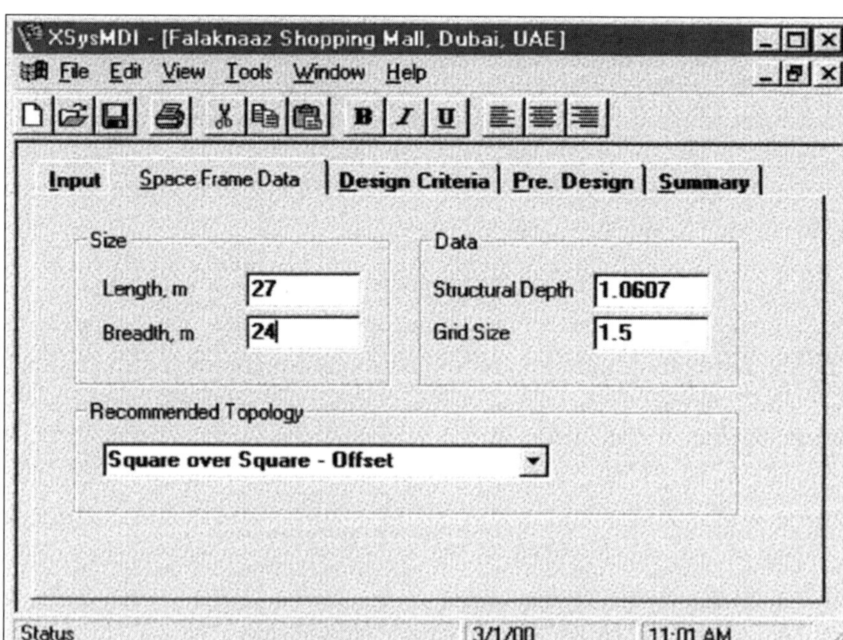


*Fig. 9.18. The CCC Xpert System Space Frame Data page*

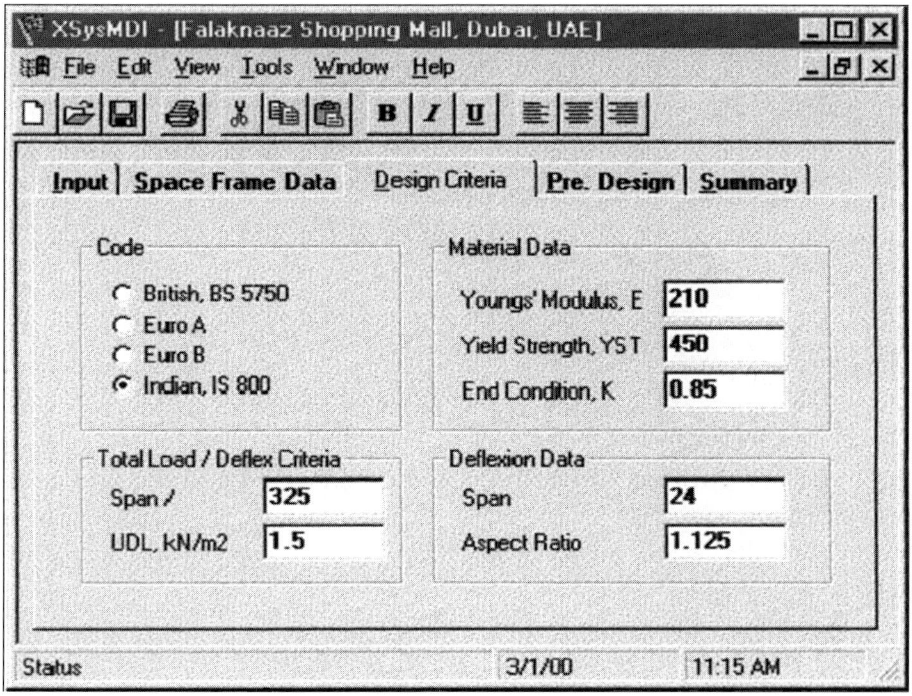


*Fig. 9.19. The CCC Xpert System Design Criteria page*

7. Move to the **Design Criteria** page, and choose the Indian Code for design from the **Code** frame. Enter the material data, such as Young's modulus $E$ (in kN/mm$^2$), the yield strength YST (in MPa) and the end-coefficient $k$ for compression members. Similarly, in the **Total Load/Deflex Criteria** frame enter the span/deflection ratio and the uniformly distribute load (UDL, in kN/m$^2$). The span and aspect ratio are automatically calculated and displayed in the **Deflection Data** frame (Fig. 9.19).
8. Now proceeded to the **Pre-Design** page. Click on the **Analyze** command button to view the results. The member lengths for the top chords and the bottom chords calculated on the basis of topology and grid size are displayed. The maximum force and deflection based on the plate analogy are displayed. The calculated maximum deflection is found to be less than the permissible deflection. However, if the deflection exceeds the permissible limit, the expert system will choose the next size. Preliminary sections for top, bottom and bracing members are displayed (Fig. 9.20).
9. Go to the **Summary** page to see the summary of the entire session (Fig. 9.21).

Finally:

- the selected members sizes are used for the first cycle of the analysis using the SAP90 program
- the customized external program MEMCO, developed by the first author, is used to optimize the space frame weight to 116 kN.

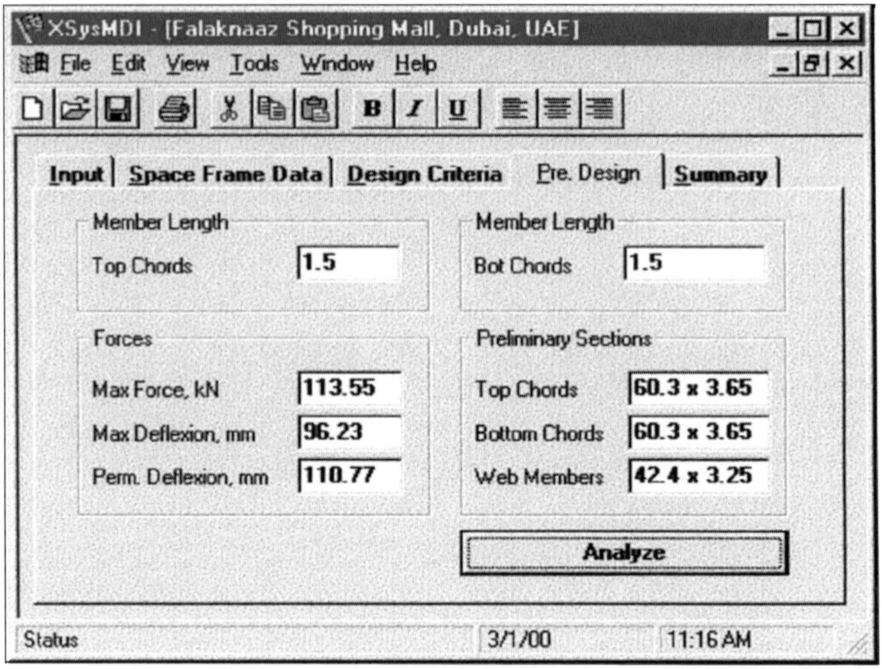


*Fig. 9.20. The CCC Xpert System Design page*

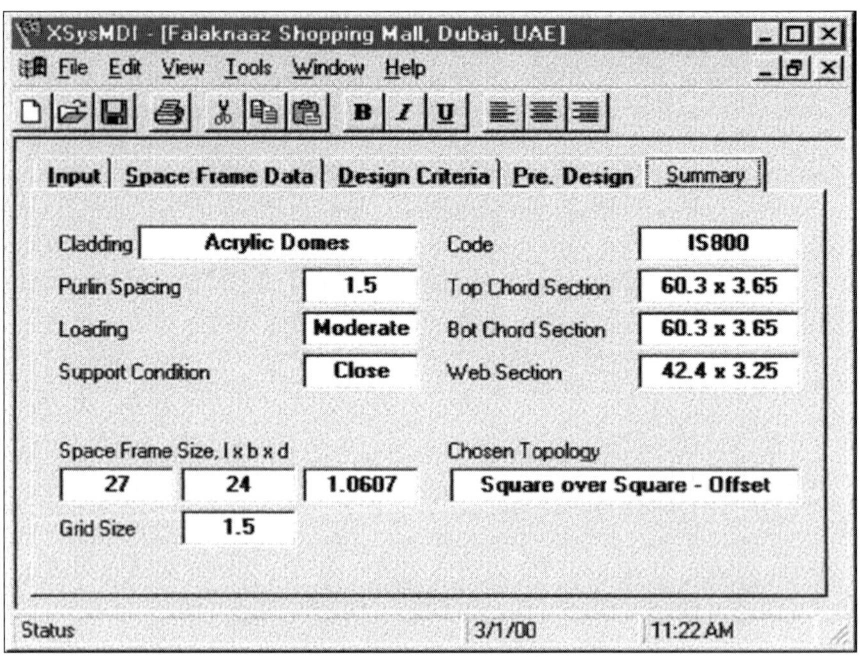


*Fig. 9.21. The CCC Xpert System Summary page*

## 9.6. References

9.1. GOLDBERG, D. E. *Genetic Algorithms in Search Optimization and Machine Learning*, Addison Wesley Longman, Reading, MA, 1999, first ISE print.
9.2. RAJEEV, S. and KRISHNAMOORTHY, C. S. Discrete optimization of structures using genetic algorithms. *Journal of Structural Engineering, ASCE*, 1997, **118**(5), 1233–1250.
9.3. GOLDBERG, E. and SAMTANI, M. P. Engineering optimization and genetic algorithm. *Proceedings of the Ninth Electronic Conference, ASCE*, New York, 1986, pp. 471–482.
9.4. THIERAUF, C. and CAI, J. Structural optimization based on evolution strategy. In: M. Papadrakakis and G. Bugeda G. (eds), *Advanced Computation Methods in Structural Analysis*, CIMNE, Barcelona, 1996, pp. 266–280.
9.5. PAPADRAKAKIS, M., LAGAROS, N. D. and TSOMPANAKIS, Y. Structural optimization using evolution strategies and neural networks. *Computer Methods in Applied Mechanics and Engineering*, 1998, **156**, 309–333.
9.6. FREEMAN, A. J. and SKAPURA, M. D. *Neural Networks*, Addison Wesley Longman, Reading, MA, 1997, p. 27, international student edition, first ISE reprint.
9.7. GUNARATNAM, D. and GERO, J. S. Effects of representation on the performance of neural networks in structural engineering applications. *Microcomputers in Civil Engineering*, 1994, **9**, 97–108.
9.8. LIU, X. and GAN, M. A preliminary structural design of expert system (SPREAD-1) based on neural networks. In: J. S. Gero (ed.), *Artificial Intelligence in Design*, Butterworth Heinemann, Oxford, 1991, pp. 785–800.
9.9. Buckingham, E. Model experiments and the form of empirical equations. *Transactions of the ASME*,1915, **37**, 263–296.
9.10. RAJEEV, S. Genetic algorithm-based methodologies for design optimization of framed structures, PhD thesis, Indian Institute of Technology, Madras, 1993.
9.11. OHMORI, H., KAWAMURA, H. and OSADA, S. Creation of space frames by genetic algorithms. In: M. Papadrakakis, A. Samartin and R. E. Onate (eds), *Proceedings of IASS–IACM Conference on Computational Methods for Shell and Spatial Structures*, ISAR, Athens, 2000.
9.12. RECHENBERG, I. *Evolution Strategy in Optimization of Technical Systems according to the Principles of Biological Evolution*. Frommann-Holzboog, Stuttgart, 1973.
9.13. SCHWEFEL, H. P. *Numerical Optimization of Computer Models*, Wiley & Sons, Chichester, 1981.
9.14. PAPADRAKAKIS, M., LAGAROS, N. D. and TSOMPANAKIS, Y. Optimization of large-scale 3D trusses using evolution strategies and neural networks. *International Journal of Space Structures*, 1999, **14**(3), 211–223.
9.15. THIERAUF, G. and CAI, J. A two-level parallel evolution strategy for solving mixed-discrete structural optimization problems. *21st ASME Design Automation Conference*, Boston, MA, 1995, p. 221.
9.16. THIERAUF, C. and CAI, J. Structural optimization based on evolution strategy. In: M. Papadrakakis and G. Bugeda (eds), *Advanced Computation Methods in Structural Analysis*, CIMNE, Barcelona, 1996, pp. 266–280.
9.17. MICHALIEWICZ, Z. *Genetic Algorithms + Data Structures – Evolution Programs*, Springer-Verlag, Berlin, 1992.
9.18. PAPADRAKAKIS, M., TSOMPANAKIS, Y. and LAGAROS, N. D. Structural shape optimization using evolution strategies. *Engineering Optimization Journal*, in press.
9.19. THIERAUF, G. and CAI, J. A two-level parallel evolution strategy for solving mixed-discrete structural optimization problems. *21st ASME Design Automation Conference*, Boston, MA, 1995, p. 221.
9.20. SCHMIT, L. A. Structural design by systematic synthesis. *Proceedings of the Second ASCE Conference on Electronic Computations*, Pittsburgh, 1960, pp. 105–122.

9.21. Papadrakakis, M, Lagaros, N. D. and Tsompanakis, Y. Structural optimization using evolution strategies and neural networks. *Computer Methods in Applied Mechanics and Engineering*, 1998, **156**, 309–333.
9.22. Joglekar, S. G., Khanra, K. K, Navani, K. P. and Dadloni, N. S. Long span space frame hangar roof with suspended docking system and cranes for heavy maintenance of 747 400 aircraft for Air India at Mumbai, India. *International Journal of Space Structures*, 1999, **14**(3), 197–209.
9.23. Feigenbaum, E. Expert systems in the 80s. In: A. Bond (ed.), *Machine Intelligence*, Infotech State of the Art Series 9, No. 3. Pergamon, Oxford, 1981.
9.24. Weiss, C. A. and Kulikowski, C. A. *A Practical Guide to Designing Expert Systems*, Rowman & Allenhead, Tatowa, NJ, 1984.
9.25. Adeli, H. (ed.). *Expert Systems in Construction and Structural Engineering*, Chapman & Hall, London, 1988.
9.26. Rule Machines Corporation. *Visual Rule Studio – Developer's Guide*, Rule Machines Corporation, 134 Fifth Avenue, Suite No. 205, Indiantic, FL 32903, April 1998.

# 10. Stability checks

## 10.1. The phenomenon of buckling

The phenomenon of buckling is best understood by considering a centrally loaded strut (Fig. 10.1(a)). Initially, the strut undergoes only axial shortening. As the load is progressively increased, a point is reached when it will deform laterally and the lateral deflection will increase significantly without any increase in the load. The load corresponding to this behaviour is termed the *critical load*. The load–deformation plot up to this point is a straight line (Fig. 10.1(b)). Linear Euler theory predicates that, on reaching the critical load, the strut may either remain straight in a state of unstable equilibrium or deform laterally (Fig. 10.1(c)).

Because of the possibility of such dual behaviour, this critical point is often referred to as a *bifurcation point*. Linear Euler theory is incapable of predicting the equilibrium path beyond the critical point. The behaviour just described is true only of ideal struts, which are initially straight with a symmetrical distribution of material and a truly concentric application of the load. These ideal conditions are seldom

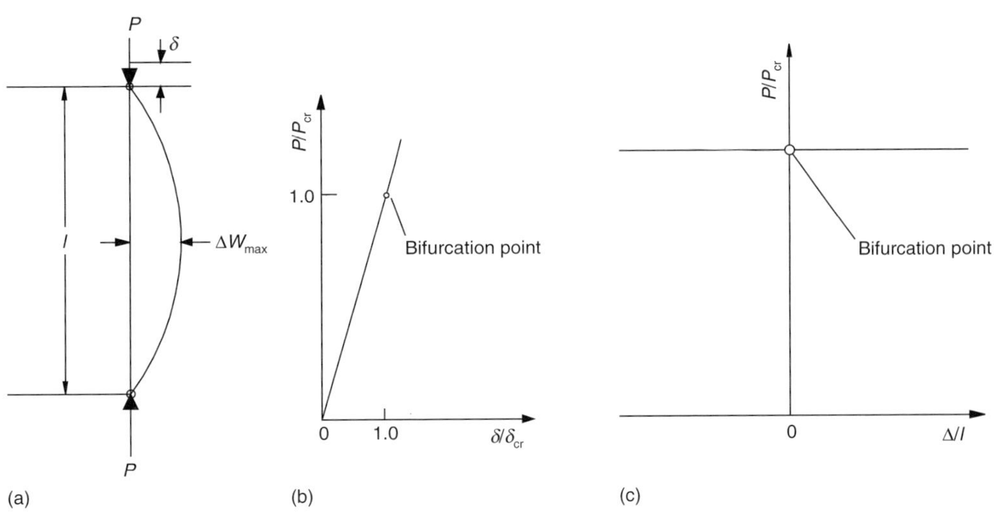


*Fig. 10.1. The buckling of an ideal elastic strut according to linear Euler theory. (Reproduced with permission from Medwadowski [10.1])*

met with in practice and hence it is necessary to consider the effect of initial imperfections (Fig. 10.2(a)). The effects of initial imperfections are depicted in Fig. 10.2(b).

In Fig. 10.2(c), the effects of geometrical non-linearity are illustrated. It can be seen that the strut is still capable of supporting a compressive load; in fact, this post-buckling load may be slightly higher than the buckling load. However, this increase is not usable in practice because it is accompanied by unacceptably large transverse deflections. What has been said of the simple strut is also true of complex structural systems.

The linear theory is not capable of predicting behaviour in the post-buckling range. A non-linear formulation is necessary. With such a formulation, the equilibrium path can be plotted in the vicinity of critical points and at post-critical points [10.1].

## 10.2. Buckling of space trusses

The types of buckling associated with space trusses may be categorized as:

- member buckling
- joint buckling
- general buckling.

In general, buckling behaviour may involve bifurcation, a limit point or snap-through (Fig. 10.3).

At a limit point, the stiffness of the structure is zero and no neighbouring equilibrium state is possible for a slight increase in load. The structure is obliged to seek a new equilibrium involving a dynamic jump, as shown in Fig. 10.3(c). The two-bar truss, discussed in Section 10.5, exhibits such behaviour.

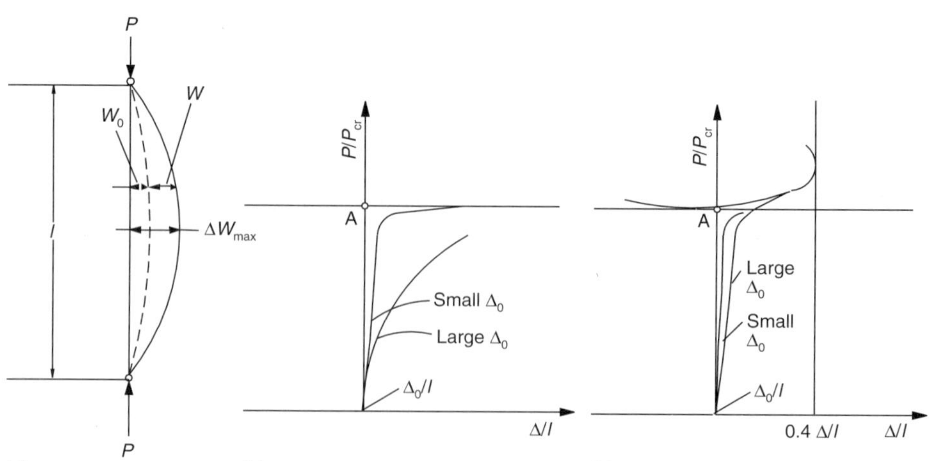


*Fig. 10.2. The effect of initial imperfections. (Reproduced with permission from Medwadowski [10.1])*

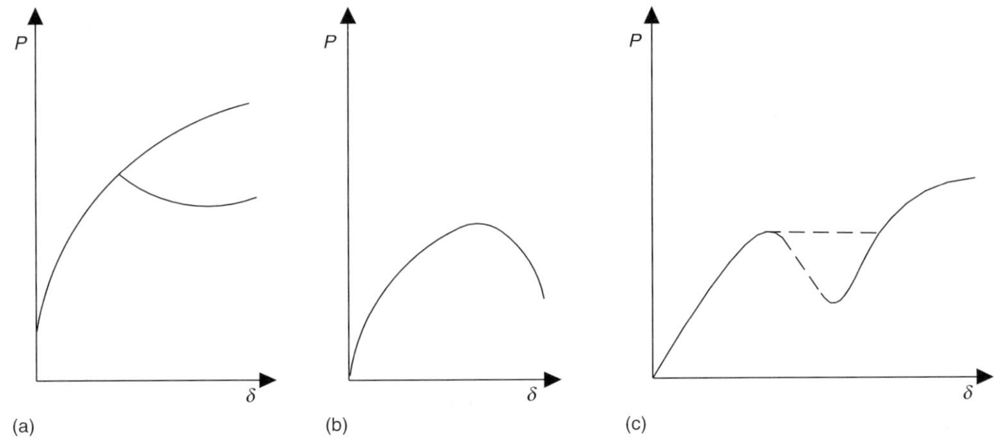


*Fig. 10.3. Different manifestations of buckling: (a) bifurcation; (b) limit point; (c) snap-through. (Reproduced with permission from Medwadowski [10.1])*

## 10.3. Discrete non-linear analysis

In the past, buckling of space frames such as reticulated domes was studied by replacing them by their continuum equivalents [10.2]. With the advent of the digital computer and the availability of powerful software packages, such as Abaqus, the investigation of the reticulated structure itself has become a rather straightforward job, once the governing differential equations have been formulated. Following reference [10.3], the major problems involved in this approach may be summed up as:

- conceptual difficulties
- correct modelling of elements and joints
- selection of numerical techniques for the solution of the set of non-linear differential equations.

In what follows a brief survey is presented of the available techniques and their relative merits.

## 10.4. Numerical methods for plotting the equilibrium path in non-linear problems

The methods available may be broadly classified as follows:

- load incremental methods
- displacement incremental methods
- arc length incremental methods
- perturbation methods
- general incremental methods.

In the load incremental method, at critical points such as bifurcation and limit points, the determinant of the coefficient matrix becomes zero and a numerically unstable situation develops in their vicinity. Hence the equilibrium path beyond the critical point cannot be traced. To overcome this problem, the displacement increment method [10.4] or the perturbation method [10.5] may be used. The disadvantage of these methods is the difficulty involved in selecting an incremental parameter in the vicinity of the critical points. The arc length method [10.6] overcomes this problem and can be used for tracing highly non-linear equilibrium paths. The drawback of the method is that a non-linear constraint concerning the arc length has to be introduced. In what follows, the constant arc length method due to Riks and Wempner [10.7] is illustrated by its application to the problem of a hinged two-bar truss which behaves non-linearly from the start.

## 10.5. Non-linear formulation of the problem of the two-bar truss

Consider a point $P$ on the member AB at a distance $a_1$ from point A of the two-bar truss (Fig. 10.4). The quantities $a_1$ and $a_2$ are local co-ordinates, with their origin at A for member AB and at B for member BC. The truss is loaded by loads $R$ and $S$ at B. B moves to B′ after undergoing deformations $u$ and $v$. The original length AB and the deformed length AB′ are denoted respectively by $d$ and $d^*$. At the point $P$ the displacements in the $a_1$ and $a_2$ directions are denoted by $\tilde{u}_1$ and $\tilde{u}_2$, respectively. Because the bar is uniformly stressed, the displacements $\tilde{u}_1$ and $\tilde{u}_2$ may be written as

$$\tilde{u}_1 = (-v\sin\theta + u\cos\theta)\frac{a_1}{d} \tag{10.1a}$$

$$\tilde{u}_2 = (-v\cos\theta - u\sin\theta)\frac{a_1}{d} \tag{10.1b}$$

For bar BC,

$$\tilde{u}_1 = (v\sin\theta + u\cos\theta)\frac{a_1}{d} \tag{10.2a}$$

$$\tilde{u}_2 = (-v\cos\theta + u\sin\theta)\frac{a_1}{d} \tag{10.2b}$$

Using Green's tensor, the strain energy for the bar AB may be written as

$$U_{\mathrm{AB}} = \frac{1}{2}\int E\left[\frac{\partial\tilde{u}_1}{\partial a_1} + \frac{1}{2}\left(\frac{\partial\tilde{u}_1}{\partial a_1}\right)^2 + \frac{1}{2}\left(\frac{\partial\tilde{u}_2}{\partial a_1}\right)^2\right]^2 \mathrm{d}V \tag{10.3}$$

A similar expression may be written for member BC.

On substituting for derivatives,

$$U = \frac{1}{2}AEd\left[\frac{-v\sin\theta + u\cos\theta}{d} + \frac{1}{2}\left(\frac{u^2}{d^2} + \frac{v^2}{d^2}\right)\right]^2 + \frac{1}{2}AEd\left[\frac{v\sin\theta + u\cos\theta}{d} + \frac{1}{2}\left(\frac{u^2}{d^2} + \frac{v^2}{d^2}\right)\right]^2 \tag{10.4}$$

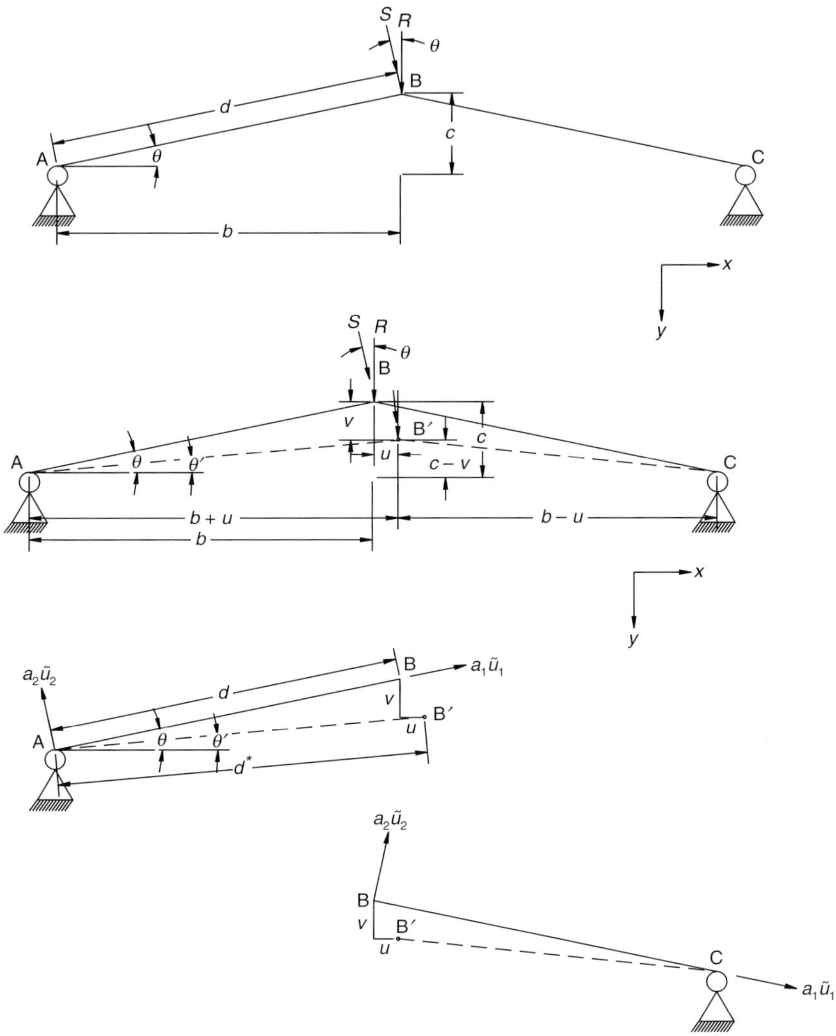


*Fig. 10.4. The undeformed (solid lines AB and BC, point ABC, angle θ) and deformed (dashed lines AB′ and B′C, point B′, angle θ′) two-bar truss*

Differentiating $U$ with respect to $u$ and $v$, and simplifying the results, we get

$$\frac{\partial U}{\partial v} = \frac{AE}{d^3}[(v-c)(v^2-2vc+u^2)] \tag{10.5}$$

$$\frac{\partial U}{\partial u} = \frac{AEu}{d^3}[(v^2+2(b^2-vc)+u^2] \tag{10.6}$$

By Castigliano's theorem, we may write:

$$f_y^{\mathrm{B}} = \frac{\partial U}{\partial v} \tag{10.7a}$$

$$f_x^{\mathrm{B}} = \frac{\partial U}{\partial u} \tag{10.7b}$$

where $f_y^{\mathrm{B}}$ and $f_x^{\mathrm{B}}$ are the components of the forces in the $v$ and $u$ directions. Assuming $R$ to be conservative (i.e. it retains its line of action) and $S$ to be non-conservative but always remains normal to AB′, we get

$$f_y^{\mathrm{B}} = R + S\cos\theta' \tag{10.8}$$

$$f_x^{\mathrm{B}} = S\sin\theta' \tag{10.9}$$

Referring to Fig. 10.4(b)

$$\cos\theta' = (b+u)/d^* \tag{10.10}$$

$$\sin\theta' = (c-v)/d^* \tag{10.11}$$

and

$$d^* = \sqrt{(b+u)^2 + (c-v)^2} \tag{10.12}$$

Combining equations (10.5) to (10.11) leads to the following equilibrium equations:

$$\frac{EA}{d^3}(v-c)(v^2 - 2vc + u^2) = R + S(b+u)/d^* \tag{10.13}$$

and

$$\frac{EAu}{d^3}[v^2 + 2(b^2 - vc) + u^2] = S(c-v)/d^* \tag{10.14}$$

### *10.5.1. Special cases*

*Case A: S = 0*

The equilibrium equations become

$$\frac{EA}{d^3}(v-c)(v^2 - 2vc + u^2) = R \tag{10.15}$$

and

$$\frac{EAu}{d^3}[v^2 + 2(b^2 - 2vc) + u^2] = 0 \tag{10.16}$$

*Case B: S = 0, u = 0*

$$\frac{EA}{d^3}(v-c)(v^2 - 2vc) = R$$

or

$$\frac{EA}{d^3}(v^3 - 3v^2c + 2vc^2) = R \tag{10.17}$$

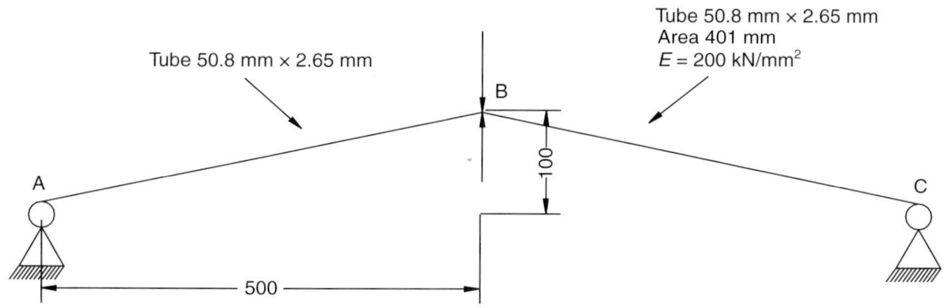


*Fig. 10.5. The two-bar pin-jointed truss considered in Example 10.1*

Substituting $\mu = c/b$, the *equilibrium equation* takes the form

$$\frac{2\mu^2 AEb^2}{d^3}v - \frac{3\mu AEb}{d^3}v^2 + \frac{AE}{d^3}v^3 = R \tag{10.18}$$

On differentiation once with respect to $v$ we arrive at the following *incremental equilibrium equation*:

$$\left(\frac{2\mu^2 AEb^2}{d^3} - \frac{6\mu AEb}{d^3}v + \frac{3AE}{d^3}v^2\right)\Delta v = \Delta R \tag{10.19}$$

Equation (10.19) may be rewritten as

$$(\mathbf{K} + \mathbf{N}_1 + \mathbf{N}_2)\Delta v = \Delta R \tag{10.20}$$

where $[\mathbf{K}]$, $[\mathbf{N}_1]$ and $[\mathbf{N}_2]$ are the Mallet and Marcal incremental matrices [10.8] redefined by Rajasekaran and Murray [10.9]. These may be written as:

$$\mathbf{K} = 2\mu^2 AEb^2/d^3 \tag{10.21}$$

$$\mathbf{N}_1 = -6\mu AEbv/d^3 \tag{10.22}$$

$$\mathbf{N}_2 = (3AE/d^3)v^2 \tag{10.23}$$

Equation (10.18) may now be rewritten as:

$$([\mathbf{K}] + \tfrac{1}{2}[\mathbf{N}_1] + \tfrac{1}{3}[\mathbf{N}_2])v = R \tag{10.24}$$

In Section 10.6.1, the matrices $[\mathbf{K}]$, $[\mathbf{N}_1]$ and $[\mathbf{N}_2]$ will be systematically derived for a space truss with multiple degrees of freedom.

### *Example 10.1*

Our task is to determine the load–deformation response for the two-bar pin-jointed truss shown in Fig. 10.5 under vertical loading at B. The data are as follows:

Area of members, $A = 401$ mm$^2$
Young's modulus, $E = 200$ kN/mm$^2$
$b = 500$ mm, $c = 100$ mm, $c/b = \mu = 0.20$.

*Solution.* First, we compute $\mathbf{K}$, $\mathbf{N}_1$ and $\mathbf{N}_2$ as follows:

$$\mathbf{K} = 2\mu^2 AEb^2/d^3 = 12.09886$$
$$\mathbf{N}_1 = -6\mu AEbv/d^3 = -0.362969876\ v$$
$$\mathbf{N}_2 = (3AE/d^3)v^2 = 0.001814829\ v^2$$

We now formulate the incremental equilibrium and equilibrium equations as follows:

$$(12.09886 - 0.362969876\ v + 0.001814829\ v^2)\Delta v = \Delta R \tag{10.25}$$

and

$$(12.09886 - 0.1814849\ v + 0.000604943\ v^2)v = R \tag{10.26}$$

The Riks–Wempner constant arc length technique involving iterative adjustments in both the loads and displacements is now used to trace the equilibrium path and arrive at the snap-through load.

*Step 1.* We choose a reference load $F$, which is usually assumed as 40% of the estimated limit load. In this problem the estimated limit load is 200 kN, and hence the reference load is chosen as $F_{11} = 80$ kN.

A load multiplication factor $\lambda_i^j$ has to be selected. The subscript and superscript denote the number of the load step and the number of the iteration within that step, respectively. Each step may involve one or two iterations before convergence is reached. Let us select $\lambda_i^j$ as 1.0. Because $v = 0$ at the beginning of this step, the tangent stiffness matrix $(\mathbf{K} + \mathbf{N}_1 + \mathbf{N}_2) = \mathbf{K}$, as $\mathbf{N}_1$ and $\mathbf{N}_2$ are zero. The tangent stiffness matrix is assumed to remain constant for all computations within a step, and it will be updated at the beginning of the next step using the current value of $v$. The displacement $(\Delta v)_1^1$ resulting from the application of the force $\lambda_1^1(80) = 1.0 \times 80 = 80$ kN may be written as

$$(\Delta v)_1^1 = \frac{80}{K} = \frac{80}{12.09886} = 6.61219 \tag{10.27}$$

Referring to Fig. 10.6, we now compute the arc length AS:

$$\text{arc length} = \sqrt{1 + 6.61219^2} = 6.68738 \tag{10.28}$$

Computing $\mathbf{K}$, $\mathbf{N}_1$ and $\mathbf{N}_2$ for $v = 6.61219$, the residual load is found to be

$$(\psi)_1^1 = 80 - \left( K + \frac{1}{2}[\mathbf{N}_1] + \frac{1}{3}[\mathbf{N}_2] \right) v = 7.759793$$

This residual force is represented in Fig. 10.6 by PS.

The displacement $(\Delta v_R)_1^2$ caused by the residual load is

$$(\Delta v_R)_1^2 = \frac{7.759793}{12.09886} = 0.64136$$

To achieve convergence let an incremental load of $\lambda^2 F_{11}$ be applied, where $\lambda_1^2$ is yet to be determined. This force will cause a displacement of $\lambda_1^2 (\Delta v)_1^1$.

Iterations are carried out along direction PQ normal to the initial tangent AP. Hence the dot product $\boldsymbol{AP} \cdot \boldsymbol{PQ} = 0$

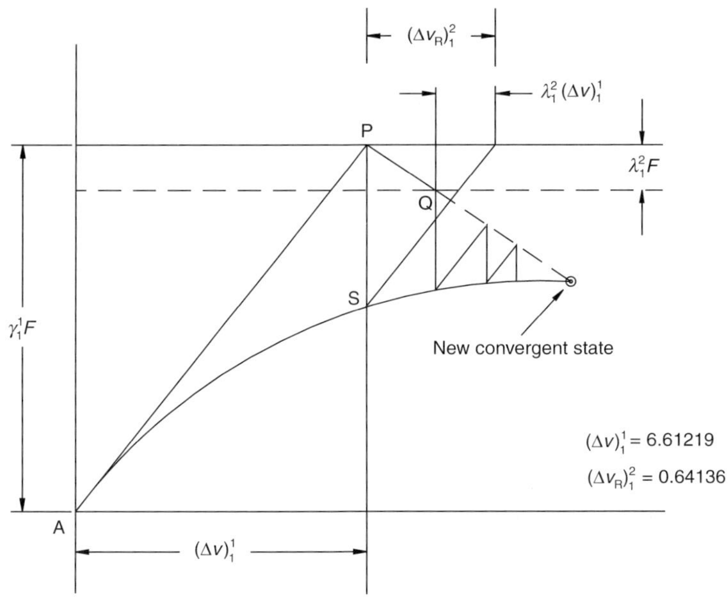


*Fig. 10.6. The Wiks–Wempner iteration for the truss considered in Example 10.1*

$$\boldsymbol{AP} = (\Delta v)_1^1 i + 1.0 j \tag{10.29}$$

$$\boldsymbol{PQ} = [(\Delta v_R)_1^2 + \lambda_1^2 (\Delta v)_1^1] i + \lambda_1^2 j \tag{10.30}$$

Hence,

$$(\Delta v)_1^1 [(\Delta v_R)_1^2 + \lambda_1^2 (\Delta v)_1^1] + \lambda_1^2 = 0 \tag{10.31}$$

Simplifying,

$$\lambda_1^2 = -\frac{(\Delta v)_1^1 (\Delta v_R)_1^2}{1 + [(\Delta v)_1^1]^2} \tag{10.32}$$

On substituting appropriate values the value of $\lambda_1^2$ is found to be –0.094827.

The values of the force and displacement at the end of the first iteration in Step 1 are:

force = (1 – 0.094827)80 = 72.4136

displacement = 6.61219(1 – 0.094827) + 0.64136 = 6.626536

The computer program RIKSWEMP, which is included on the CD-ROM that accompanies this book, can be used to carry out the computations involved in the first step. We use the following input data for this problem:

$A = 400$, $E = 200$, $B = 500$, $C = 100$, $T = 0.20$, $F_{11} = 80$, $\text{LAM}_1 = 1.0$

The program gives the force and displacement at the end of the first step as

$F_{12} = 72.41376$ and $V_{12} = 6.626536$, respectively. It also gives the initial tangent modulus $K_T$ (9.773343), the arc length (6.687385) and the appropriate $LAM_{21}$ (0.9025647) with which to commence step 2.

The CD-ROM also contains the program NONLIN, which computes the residual force at the end of step 1. The input data are

$A = 401, B = 500, C = 100, T = 0.20, V = 6.626536, F = 72.41376$

The program calculates the residual force as 0.0330426 kg. This residual force is very small, and so convergence may be assumed and no additional iterations are necessary.

The program gives the converged force and displacement at the end of the first step as $F_{12} = 72.41376$ and $V_{12} = 6.626536$, the initial tangent modulus $K_T$ (9.773343), the arc length (6.687385) and the appropriate $LAM_{21}$ (0.9025647) to commence step 2.

*Step 2.* Starting with step 2, the program WEMP, which is on the accompanying CD-ROM, can be used to obtain the converged load and displacement at the end of each step. The input data are:

$A = 401, E = 200, B = 500, C = 100, T = 0.20, F = 80, F_{11} = 72.41376, V_1 = 6.626536,$
$LAM_{11} = 0.9025647$

The program gives the converged force and displacement at the end of step 2 as $F_{13} = 154.7461$ and $V_{13} = 16.77138$. It also gives the value of $LAM_{21}$ (0.4708042) to be used as $LAM_{11}$ to commence step 3. The residual force is found from NONLIN by inputting: 401, 200, 500, 100, 0.20, 16.77138 and 154.7641. The residual force is calculated to be $4.312684 \times 10^{-2}$.

*Step 3.* Proceeding in a similar manner to above, WEMP is used to calculate the force and displacement at the end of step 3 as 203.6187 and 26.26799. The program gives the value of $LAM_{21}$ for use as $LAM_{11}$ in step 4 as 0.1877275. NONLIN calculates the residual force to be $6.586139 \times 10^{-2}$.

*Step 4.* At the end of step 4, WEMP gives the converged force as 230.7446 and the displacement as 37.74726. The value of $LAM_{11}$ suggested for the next step is $6.715479 \times 10^{-2}$.

*Step 5.* When WEMP is used for this step, the program signals on overflow in $\mathbf{N}_2$. This is an indication that we are very close to the limit load.

*Step 6.* We go back to end of step 3 and run WEMP with the following input data:

$A = 401, E = 200, B = 500, C = 100, T = 0.20, F = 80, F_1 = 203.6187, V_1 = 26.26799,$
$LAM_{11} = 0.2025$

The program gives the load and displacement as 232.9002 and 41.34783, respectively. NONLIN gives the residual force as 0.1461769. This value is as close to the limit load as it is possible to get. The snap-through load may be estimated as 232.9002 kN, without appreciable error. The load displacement path is shown in Fig. 10.7.

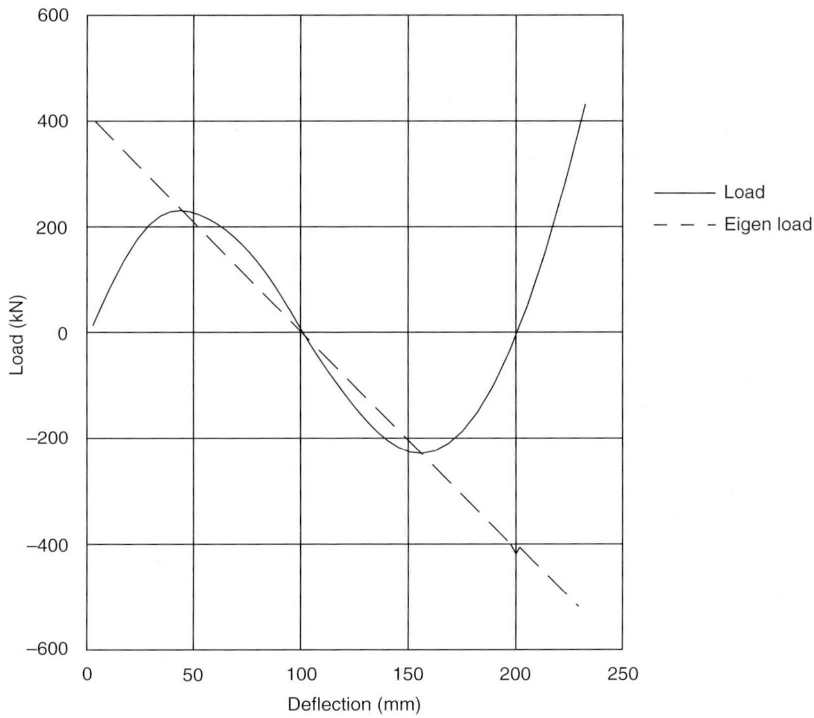


*Fig. 10.7. The equilibrium path for the case considered in Example 10.1*

*Checking the snap-through load*

In this problem it is, fortunately, possible to plot the load–displacement path directly from the cubic equation

$$0.000604943v^3 - 0.181485v^2 + 12.09886v = R \tag{10.33}$$

$$\frac{\mathrm{d}R}{\mathrm{d}v} = v^2 - 200v + 667$$

Setting equation (10.33) to zero, we arrive at the result that $R = 0$ at $v = 100$ (Fig. 10.8).

Setting $\mathrm{d}R/\mathrm{d}v = 0$ and solving for $v$, the snap-through load occurs at $v = 42.27$. Now, from equation (10.33), $R$ is computed as 232.84. This is in close agreement with the result of 232.9002 found from the Riks–Wempner iteration.

*Computation of the eigen load*

The stiffness matrix of the structure may be regarded as the sum of $[\mathbf{K}]$ and $[\mathbf{K}_G]$. Referring to equation (10.20), the critical condition is reached when the stiffness matrix is singular. This requires that $\det([\mathbf{K}] + [\mathbf{K}_G]) = 0$, or

$$\det\{[\mathbf{K}] + \lambda([\mathbf{N}_1] + [\mathbf{N}_2])\} = 0 \tag{10.34}$$

Corresponding to each equilibrium load $L$, the eigen load is found as

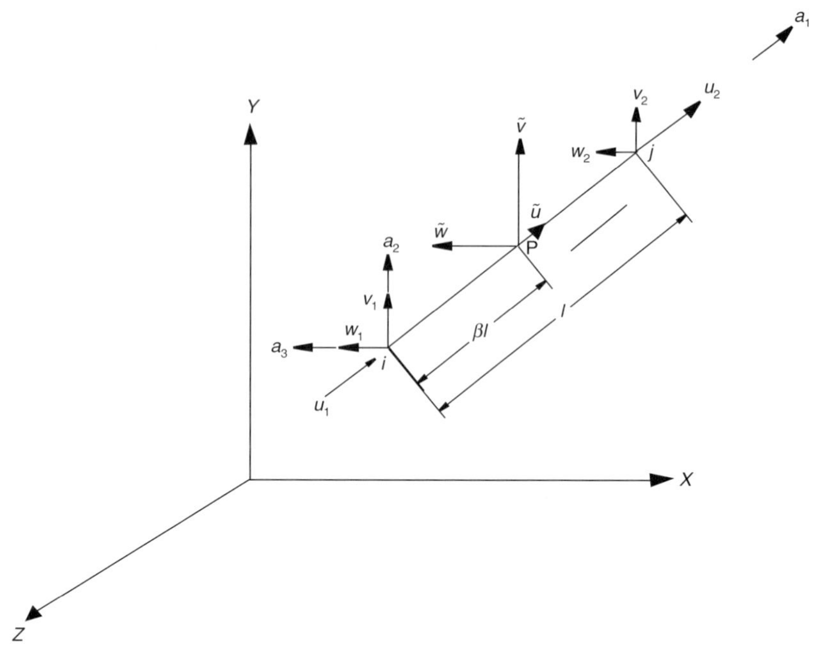


*Fig. 10.8. The local co-ordinates and displacements of the bar considered in Example 10.1*

$$\text{eigen load} = -\frac{[K]L}{[N_1]+[N_2]} \tag{10.35}$$

For example, at the end of step 1 we arrived at an equilibrium load of 72.41376 with a corresponding displacement of 6.626536. The eigen load is computed by using equation (10.35) in the program NONLIN. The program calculates the eigen load as 376.744 kN.

It is interesting to note that the eigen load corresponding to the snap-through load is the snap-through load itself, because at this point on the load–displacement path $\mathbf{K} = \mathbf{N}_1 + \mathbf{N}_2$. The variation in the eigen load with the displacement turns out to be a straight line (see Fig. 10.7). It may be noted that the eigen load is an upper bound on the snap-through load.

## 10.6. Non-linear formulation of the stability of space trusses with multiple degrees of freedom

A bar forming part of a pin-jointed truss is shown in space in Fig. 10.8, with its ends designated as $i$ and $j$. Assuming that the origin is at $i$, local co-ordinates $a_1$, $a_2$ and $a_3$ are defined such that $a_1$ lies along the bar and $a_2$ and $a_3$ are orthogonal to it. Let the displacements at an arbitrary point on the bar at $\beta l$ from $i$ be $\tilde{u}$, $\tilde{v}$ and $\tilde{w}$. The parameter $l$ is the length of the bar. The bar has six degrees of freedom and its displacements are $u_1$, $v_1$ and $w_1$ at $i$ and $u_2$, $v_2$ and $w_2$ at $j$. For uniform tension, we may write

$$\begin{aligned}\tilde{u} &= u_1(1-\beta)+u_2\beta \\ \tilde{v} &= v_1(1-\beta)+v_2\beta \\ \tilde{w} &= w_1(1-\beta)+w_2\beta\end{aligned} \tag{10.36}$$

The strain energy of the bar by Green's theorem is

$$U = ½EAl\varepsilon^2$$

where

$$\varepsilon = \frac{\partial \tilde{u}}{\partial a_1} + \frac{1}{2}\left[\left(\frac{\partial \tilde{u}}{\partial a_1}\right)^2 + \left(\frac{\partial \tilde{v}}{\partial a_1}\right)^2 + \left(\frac{\partial \tilde{w}}{\partial a_1}\right)^2\right] \tag{10.37}$$

Substituting for $\tilde{u}$, $\tilde{v}$ and $\tilde{w}$ from equation (10.36),

$$\varepsilon = \frac{u_2 - u_1}{l} + \frac{1}{2l^2}[(u_2-u_1)^2 + (v_2-v_1)^2 + (w_2-w_1)^2] \tag{10.38}$$

We may write:

$$\varepsilon = \varepsilon_{\mathrm{L}} + \varepsilon_{\mathrm{nL}} \tag{10.39}$$

where $\varepsilon_{\mathrm{L}}$ and $\varepsilon_{\mathrm{nL}}$ are the linear and non-linear components of the total strain, respectively. Denoting the displacements by $\{\mathbf{q}\}$, we have

$$\mathbf{q}^{\mathrm{T}} = [u_1, v_1, w_1, u_2, v_2, w_2] \tag{10.40}$$

The linear component of the strain $\varepsilon_{\mathrm{L}}$ may be written in matrix form as

$$\varepsilon_{\mathrm{L}} = \mathbf{L}^{\mathrm{T}}\mathbf{q} = \mathbf{q}^{\mathrm{T}}\mathbf{L} \tag{10.41}$$

where

$$\mathbf{L}^{\mathrm{T}} = [-1 \quad 0 \quad 0 \quad 1 \quad 0 \quad 0] \tag{10.42}$$

Similarly, $\varepsilon_{\mathrm{nL}}$ may be written as

$$\varepsilon_{\mathrm{nL}} = ½\mathbf{q}^{\mathrm{T}}\mathbf{H}\mathbf{q} \tag{10.43}$$

where

$$H = \frac{1}{l^2}\begin{bmatrix} 1 & 0 & 0 & -1 & 0 & 0 \\ 0 & 1 & 0 & 0 & -1 & 0 \\ 0 & 0 & 1 & 0 & 0 & -1 \\ -1 & 0 & 0 & 1 & 0 & 0 \\ 0 & -1 & 0 & 0 & 1 & 0 \\ 0 & 0 & -1 & 0 & 0 & 1 \end{bmatrix} \tag{10.44}$$

The potential energy $\pi$ of the bar may be written as

$$\pi = U + V = U - \mathbf{q}^{\mathrm{T}}\boldsymbol{Q} \tag{10.45}$$

where $Q$ is the vector of the nodal forces in the $a_1$, $a_2$ and $a_3$ local co-ordinate system.

Substituting the expression for $\varepsilon$ from equation (10.39) in the expression for strain energy

$$U = \tfrac{1}{2} EAl\varepsilon^2 \tag{10.46}$$

we arrive at

$$\begin{aligned}\pi &= \frac{EAl}{2}\left(\mathbf{q}^{\mathrm{T}}\mathbf{L} + \frac{\mathbf{q}^{\mathrm{T}}\mathbf{H}\mathbf{q}}{2}\right)\left(\mathbf{L}^{\mathrm{T}}\mathbf{q} + \frac{\mathbf{q}^{\mathrm{T}}\mathbf{H}\mathbf{q}}{2}\right) - \mathbf{q}^{\mathrm{T}}Q \\ &= \frac{EAl}{2}\left(\mathbf{q}^{\mathrm{T}}\mathbf{L}\mathbf{L}^{\mathrm{T}}\mathbf{q} + \frac{1}{2}\mathbf{q}^{\mathrm{T}}(\mathbf{L}\mathbf{q}^{\mathrm{T}}\mathbf{H} + \mathbf{H}\mathbf{q}\mathbf{L}^{\mathrm{T}})\mathbf{q} + \frac{\mathbf{q}^{\mathrm{T}}\mathbf{H}\mathbf{q}\mathbf{q}^{\mathrm{T}}\mathbf{H}\mathbf{q}}{4}\right)\end{aligned} \tag{10.47}$$

It may be noted that $\mathbf{L}^{\mathrm{T}}\mathbf{q} = \mathbf{q}^{\mathrm{T}}\mathbf{L}$ and $\mathbf{q}^{\mathrm{T}}\mathbf{H}\mathbf{q}$ are scalar and $\mathbf{H}$ is symmetric. These scalar forms may be commuted, transposed and inserted at arbitrary locations within other matrices. Hence equation (10.47) may be rewritten as

$$\begin{aligned}\pi = \frac{EAl}{2}\Bigg[&\mathbf{q}^{\mathrm{T}}\mathbf{L}\mathbf{L}^{\mathrm{T}} + \frac{1}{3}\mathbf{q}^{\mathrm{T}}(\mathbf{L}\mathbf{q}^{\mathrm{T}}\mathbf{H} + \mathbf{L}^{\mathrm{T}}\mathbf{q}\mathbf{H} + \mathbf{H}\mathbf{q}\mathbf{L}^{\mathrm{T}})\mathbf{q} + \\ &\frac{1}{6}\mathbf{q}^{\mathrm{T}}\left(\frac{\mathbf{q}^{\mathrm{T}}\mathbf{H}\mathbf{q}\mathbf{H}}{2} + \mathbf{H}\mathbf{q}\mathbf{q}^{\mathrm{T}}\mathbf{H}\right)\mathbf{q}\Bigg] - \mathbf{q}^{\mathrm{T}}Q\end{aligned} \tag{10.48}$$

Differentiating equation (10.48) with respect to the displacement co-ordinates leads to the *equilibrium equation*, which may be rewritten as

$$\begin{aligned}\frac{\partial \pi}{\partial q_i} = EAl\Bigg[&\mathbf{L}\mathbf{L}^{\mathrm{T}}\mathbf{q} + \frac{1}{2}(\mathbf{L}\mathbf{q}^{\mathrm{T}}\mathbf{H} + \mathbf{L}^{\mathrm{T}}\mathbf{q}\mathbf{H} + \mathbf{H}\mathbf{q}\mathbf{L}^{\mathrm{T}})\mathbf{q} + \\ &\frac{1}{3}\left(\frac{\mathbf{q}^{\mathrm{T}}\mathbf{H}\mathbf{q}\mathbf{H}}{2} + \mathbf{H}\mathbf{q}\mathbf{q}^{\mathrm{T}}\mathbf{H}\right)\mathbf{q}\Bigg] - Q = \{0\}\end{aligned} \tag{10.49}$$

The *incremental equilibrium* equation is next found as

$$\begin{aligned}\frac{\partial \pi}{\partial q_i\, \partial q_j} = EAl\Bigg[&\mathbf{L}\mathbf{L}^{\mathrm{T}} + (\mathbf{L}\mathbf{q}^{\mathrm{T}}\mathbf{H} + \mathbf{L}^{\mathrm{T}}\mathbf{q}\mathbf{H} + \mathbf{H}\mathbf{q}\mathbf{L}^{\mathrm{T}}) + \\ &\left(\frac{\mathbf{q}^{\mathrm{T}}\mathbf{H}\mathbf{q}\mathbf{H}}{2} + \mathbf{H}\mathbf{q}\mathbf{q}^{\mathrm{T}}\mathbf{H}\right)\Bigg]\Delta q = \Delta Q\end{aligned} \tag{10.50}$$

Equations (10.48), (10.49) and (10.50) due to Rajasekaran and Murray [10.9] will now be recast in the Mallet and Marcal format [10.8]. Let us denote

$$[\mathbf{K}]^{\mathrm{e}} = EAl\mathbf{L}\mathbf{L}^{\mathrm{T}} \tag{10.51a}$$

$$[\mathbf{N}_1]^{\mathrm{e}} = EAl(\mathbf{L}\mathbf{q}^{\mathrm{T}}\mathbf{H} + \mathbf{L}^{\mathrm{T}}\mathbf{q}\mathbf{H} + \mathbf{H}\mathbf{q}\mathbf{L}^{\mathrm{T}}) \tag{10.51b}$$

$$[\mathbf{N}_2]^{\mathrm{e}} = EAl\left(\frac{\mathbf{q}^{\mathrm{T}}\mathbf{H}\mathbf{q}}{2} + \mathbf{H}\mathbf{q}\mathbf{q}^{\mathrm{T}}\mathbf{H}\right) \tag{10.51c}$$

Using the above notation the equations for potential energy, equilibrium and incremental equilibrium are recast in the Mallet and Marcal standard nomenclature as

$$\pi = \mathbf{q}^{\mathrm{T}}(\tfrac{1}{2}K^{\mathrm{e}} + \tfrac{1}{6}N_1^{\mathrm{e}} + \tfrac{1}{12}N_2^{\mathrm{e}})\boldsymbol{q} = \mathbf{q}^{\mathrm{T}}\boldsymbol{Q} \tag{10.52}$$

$$\frac{\partial \pi}{\partial \boldsymbol{q}_i} = (K^{\mathrm{e}} + \tfrac{1}{2}N_1^{\mathrm{e}} + \tfrac{1}{3}N_2^{\mathrm{e}})\boldsymbol{q} = \{\boldsymbol{Q}\} \tag{10.53}$$

and

$$\frac{\partial \pi}{\partial \boldsymbol{q}_i\,\partial \boldsymbol{q}_j} = (K^{\mathrm{e}} + N_1^{\mathrm{e}} + N_2^{\mathrm{e}})\Delta\boldsymbol{q} = \{\Delta\boldsymbol{Q}\} \tag{10.54}$$

The superscript e denotes that these relate to the element. $K^{\mathrm{e}}$, $N_1^{\mathrm{e}}$ and $N_2^{\mathrm{e}}$ are known as *incremental matrices*. They are symmetric and repeat themselves in the energy, equilibrium and incremental equilibrium equations. Expanding the relationships established in equations (10.52) to (10.54) using the definitions of **q**, **L** and **H**, we arrive at

$$[K]^{\mathrm{e}} = \frac{EA}{l}\begin{bmatrix} 1 & 0 & 0 & -1 & 0 & 0 \\ 0 & 0 & 0 & 0 & 0 & 0 \\ 0 & 0 & 0 & 0 & 0 & 0 \\ -1 & 0 & 0 & 1 & 0 & 0 \\ 0 & 0 & 0 & 0 & 0 & 0 \\ 0 & 0 & 0 & 0 & 0 & 0 \end{bmatrix} \tag{10.55}$$

$$[N_1^{\mathrm{e}}] = \frac{EA}{l^2}\begin{bmatrix} M & -M \\ -M^{\mathrm{T}} & M \end{bmatrix}$$

where

$$M = \begin{bmatrix} 3u_{21} & v_{21} & w_{21} \\ v_{21} & v_{21} & 0 \\ w_{21} & 0 & v_{21} \end{bmatrix} \tag{10.56}$$

where $u_{21} = u_2 - u_1$, $v_{21} = v_2 - v_1$ and $w_{21} = w_2 - w_1$.

$$[N_1^{\mathrm{e}}] = \frac{EA}{l^2}\begin{bmatrix} N & -N \\ -N^{\mathrm{T}} & N \end{bmatrix}$$

where

$$N = \begin{bmatrix} u_{21}^2 + \tfrac{1}{2}\Delta & u_{21}v_{21} & u_{21}w_{21} \\ & v_{21}^2 + \tfrac{1}{2}\Delta & v_{21}w_{21} \\ \text{symmetric} & & w_{21}^2 + \tfrac{1}{2}\Delta \end{bmatrix} \tag{10.57}$$

where $\Delta = u_{21}^{\,2} + v_{21}^{\,2} + w_{21}^{\,2}$.

All the matrices associated with a single element for a total or linear incremental solution have now been developed, and all that is now left is to assemble the equilibrium equations for the entire structure.

### 10.6.1. Transformation from local to global co-ordinates

The transformation from local co-ordinates $(a_1, a_2, a_3)$ to global co-ordinates $(X, Y, Z)$ may be accomplished as follows:

$$\begin{Bmatrix} a_1 \\ a_2 \\ a_3 \end{Bmatrix} = \begin{bmatrix} a_{11} & a_{12} & a_{13} \\ a_{21} & a_{22} & a_{23} \\ a_{31} & a_{32} & a_{33} \end{bmatrix} \begin{Bmatrix} X \\ Y \\ Z \end{Bmatrix} = [\mathbf{A}] \begin{Bmatrix} X \\ Y \\ Z \end{Bmatrix} \tag{10.58}$$

Regarding $\boldsymbol{q}$ as the assembled vector of global displacements and $\boldsymbol{Q}$ as the assembled vector of global forces and

$$\begin{aligned} [\mathbf{K}] &= \sum_{m=1}^{N} [\mathbf{T}]^{\mathrm{T}} [\mathbf{K}^{\mathrm{e}}][\mathbf{T}] \\ [\mathbf{N}_1] &= \sum_{m=1}^{N} [\mathbf{T}]^{\mathrm{T}} [\mathbf{N}_1^{\mathrm{e}}][\mathbf{T}] \\ [\mathbf{N}_2] &= \sum_{m=1}^{N} [\mathbf{T}]^{\mathrm{T}} [\mathbf{N}_2^{\mathrm{e}}][\mathbf{T}] \end{aligned} \tag{10.59}$$

and

$$\mathbf{T} = \begin{bmatrix} [\mathbf{A}] & 0 \\ 0 & [\mathbf{A}] \end{bmatrix} \tag{10.60}$$

The global equilibrium equations take the form

$$([\mathbf{K}] + \tfrac{1}{2}[\mathbf{N}_1] + \tfrac{1}{3}[\mathbf{N}_2])\{\boldsymbol{q}\} = \{\boldsymbol{Q}\} \tag{10.61}$$

$$\{[\mathbf{K}] + [\mathbf{N}_1] + [\mathbf{N}_2]\}\{\Delta \boldsymbol{q}\} = \{\Delta \boldsymbol{Q}\} \tag{10.62}$$

It now remains to derive [**A**].

### 10.6.2. Derivation of matrix [A]

[A] is the transformation matrix that converts global co-ordinates into local co-ordinates. Consider a member IJ (Fig. 10.9). Let $(X_i, Y_i, Z_i)$ and $(X_j, Y_j, Z_j)$ be the Cartesian co-ordinates of I and J referred to the global system of co-ordinates.

Adopting the notation $X_{ji} = X_j - X_i$, we may write

$$\boldsymbol{IJ} = X_{ji}\boldsymbol{i} + Y_{ji}\boldsymbol{j} + Z_{ji}\boldsymbol{k} \tag{10.63}$$

The length of the member IJ is:

$$l_{ij} = \sqrt{X_{ji}^2 + Y_{ji}^2 + Z_{ji}^2} \tag{10.64}$$

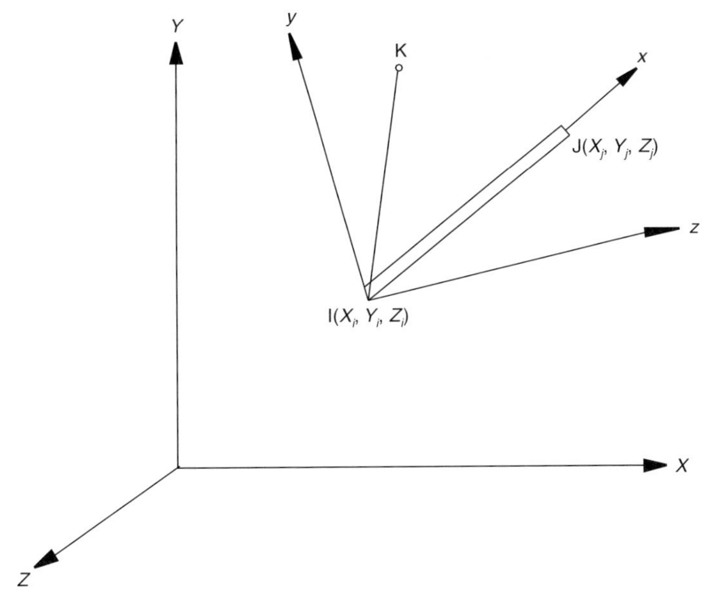


*Fig. 10.9. The local and global co-ordinates of the member IJ used in the derivation of the matrix* **[A]**

Hence the unit vector in the local $x$ direction is

$$\boldsymbol{a}_1 = a_{11}i + a_{12}j + a_{13}k \tag{10.65}$$

where

$$a_{11} = \frac{X_{ji}}{l_{ji}}, \qquad a_{12} = \frac{Y_{ji}}{l_{ji}}, \qquad a_{13} = \frac{Z_{ji}}{l_{ji}} \tag{10.66}$$

Our aim now is to find the local $y$ and $z$ directions. Consider a point K in the local $x, y$ plane:

$$\boldsymbol{IK} = X_{ki}i + Y_{ki}j + Z_{ki}k \tag{10.67}$$

The length $\boldsymbol{IK}$ is

$$l_{ik} = \sqrt{X_{ki}^2 + Y_{ki}^2 + Z_{ki}^2} \tag{10.68}$$

and the unit vector in the direction IK is given by

$$\boldsymbol{e}_{ik} = \frac{X_{ki}}{l_{ik}}i + \frac{Y_{ki}}{l_{ik}}j + \frac{Z_{ki}}{l_{ik}} \tag{10.69}$$

To find the vector perpendicular to the $i, j, k$ plane parallel to the $z$ axis, we form the vector cross-product:

$$\boldsymbol{B} = \boldsymbol{IJ} \times \boldsymbol{IK} \qquad \text{or} \qquad \boldsymbol{a}_1 \times \boldsymbol{e}_{ik}$$

$$\boldsymbol{B} = \boldsymbol{a}_1 \times \boldsymbol{e}_{ik} \begin{bmatrix} i & j & k \\ a_{11} & a_{12} & a_{13} \\ \dfrac{X_{ki}}{\lambda_{ik}} & \dfrac{Y_{ki}}{\lambda_{ik}} & \dfrac{Z_{ki}}{\lambda_{ik}} \end{bmatrix}$$

$$\boldsymbol{B} = i\left(\frac{a_{12}Z_{ki}}{l_{ik}} - \frac{a_{13}Y_{ki}}{l_{ik}}\right) - j\left(\frac{a_{11}Z_{ki}}{l_{ik}} - \frac{a_{13}X_{ki}}{l_{ik}}\right) + k\left(\frac{a_{11}Y_{ki}}{l_{ik}} - \frac{a_{12}X_{ki}}{l_{ik}}\right) \tag{10.70}$$

$$\boldsymbol{B} = B_1 i + B_2 j + B_3 k \tag{10.71}$$

$B_1$, $B_2$ and $B_3$ are readily identified by comparison with equation (10.70). As $\boldsymbol{B}$ is parallel to the local $z$ axis, the unit vector in the local $z$ axis may now be written as

$$\boldsymbol{a}_3 = a_{31}i + a_{32}j + a_{33}k \tag{10.72}$$

where

$$a_{31} = B_1/\sqrt{B_1^2 + B_2^2 + B_3^2}$$
$$a_{32} = B_2/\sqrt{B_1^2 + B_2^2 + B_3^2}$$
$$a_{33} = B_3/\sqrt{B_1^2 + B_2^2 + B_3^2}$$

We now proceed to find the vector parallel to the local $y$ axis by forming the vector product

$$\begin{aligned} \boldsymbol{c} &= \boldsymbol{a}_3 \times \boldsymbol{a}_1 \\ &= \begin{bmatrix} i & j & k \\ a_{31} & a_{32} & a_{33} \\ a_{11} & a_{12} & a_{13} \end{bmatrix} \\ &= i(a_{32}a_{13} - a_{12}a_{33}) - j(a_{31}a_{13} - a_{11}a_{33}) + k(a_{31}a_{12} - a_{11}a_{32}) \end{aligned} \tag{10.73}$$

or

$$\boldsymbol{C} = C_1 i + C_2 j + C_3 k \tag{10.74}$$

$C_1$, $C_2$ and $C_3$ may be found by comparison with equation (10.73). The unit vector $\boldsymbol{a}_2$ along y axis is found as

$$\boldsymbol{a}_2 = \frac{\boldsymbol{C}}{|\boldsymbol{C}|} = a_{21}i + a_{22}j + a_{33}k \tag{10.75}$$

where

$$\begin{aligned} a_{21} &= C_1/\sqrt{C_1^2 + C_2^2 + C_3^2} \\ a_{22} &= C_2/\sqrt{C_1^2 + C_2^2 + C_3^2} \\ a_{23} &= C_3/\sqrt{C_1^2 + C_2^2 + C_3^2} \end{aligned} \tag{10.76}$$

$$[\mathbf{A}] = \begin{bmatrix} a_{11} & a_{12} & a_{13} \\ a_{21} & a_{22} & a_{23} \\ a_{31} & a_{32} & a_{33} \end{bmatrix} \tag{10.77}$$

Several examples of space trusses with multiple degrees of freedom have been discussed by Rajasekaran [10.10]

## 10.7. Stability analysis of single-layer reticulated domes

A number of studies that have been published recently [10.11–10.14] have shed some light on the grey area relating to the stability analysis of single-layer reticulated domes. The techniques presented in these papers can take into account both geometrical and material non-linearities and sketch the equilibrium path past critical points. A few limited tests carried out on laboratory models seem to have validated these analytical predictions [10.15]. However, this activity at the research level has not been matched by field measurements of displacements on prototype structures. The first author of this book had the rare opportunity of carrying out such a full-scale test on a single-layer reticulated dome and taking deflection measurements at the test site. Dr Albermani, the co-author of this chapter, carried out a stability analysis of the pilot and final tests. In the case study that follows the analytically predicted deflections are compared with the results of non-linear analysis and the findings reached are documented.

## 10.8. Case history: Pragati Maidan domes, New Delhi, India

A cluster of 16 spherical reticulated domes of tubular steel 22.10 m × 22.10 m, on square ground plans, supported on four boundary arches, had been proposed by the architect Joseph Allen Stein, of Stein, Doshi and Bhalla, New Delhi, for roofing an Exhibition Centre at Pragati Maidan, New Delhi. The client was the Trade Fair Authority of India.

Nagarjuna Coated Tubes Limited (NCTL), Hyderabad, was contracted to supply the structural design and fabricate and erect the domes. As originally conceived, the domes of 22 m radius were to be of welded construction with nine latitudinal rings to suit the roof cladding of wood wool slabs covered with 4 cm thick mesh-reinforced concrete. The design load specified was 170 kg/m$^2$. The contractor was required to test a prototype dome submitted to the design load. A prototype dome that was test loaded by the contractor collapsed before the working load was reached. Dr Eekhout and the first author, who happened to be in New Delhi at that time, were invited by NCTL to visit the site to make an assessment of the collapse. They advised the client that site welding, which is subject to wide variations in quality, was an unsuitable technique for fabricating single-layer reticulated domes, which are particularly prone to snap-through buckling, this often being triggered by initial imperfections. We proposed that the welded design be replaced by a pin-jointed dome with node connectors resting on shop-welded boundary arches. Eekhout offered to send a more detailed report from Delft listing other changes that may be necessary.

### 10.8.1. Diagnosis of the collapse

The radius of the collapsed dome was 22 m and the members were about 1 m long. Wright's formula [10.2] for the lower and upper-bound snap-through load may be written as

$$\frac{AEl}{12R^3} < q_{cr} < \frac{AEl}{6R^3}$$

For a tube of 76.10 mm × 3.65 mm, of area 820 mm$^2$, the lower bound value of the snap-through load $q_{cr}$ is readily computed as

$$q_{cr} = \frac{820 \times 210 \times 1}{12 \times 22^3} = 1.34 \text{ kN/m}^2$$

The specified working load being 170 kg/m$^2$, it is not surprising that the test dome collapsed before the working load was reached.

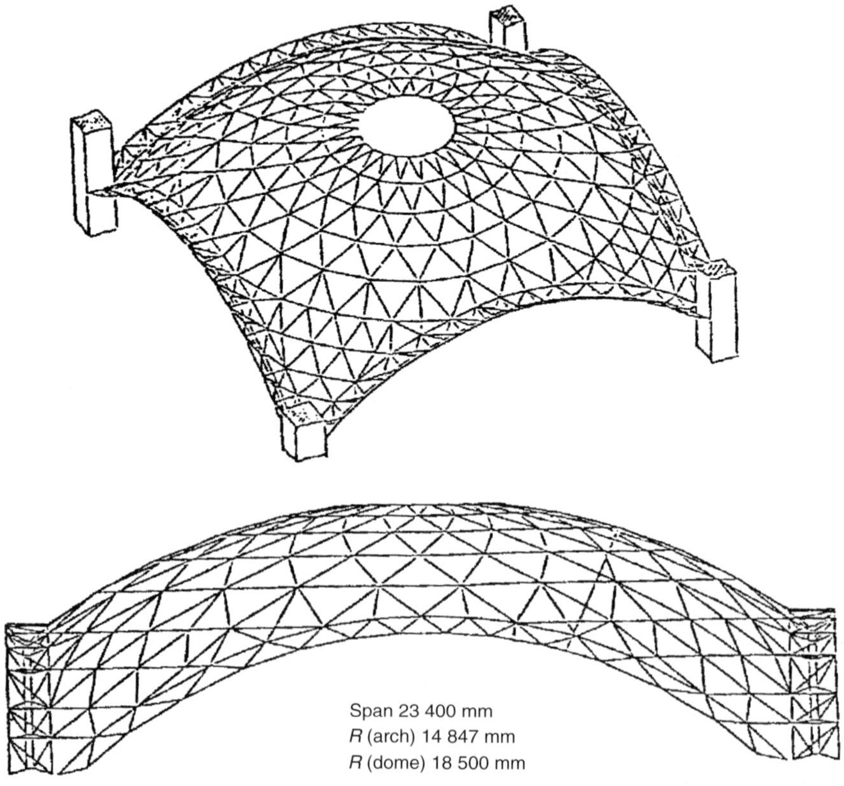


*Fig. 10.10. Three-dimensional view of the revised design for the Pragati Maidan domes, New Delhi*

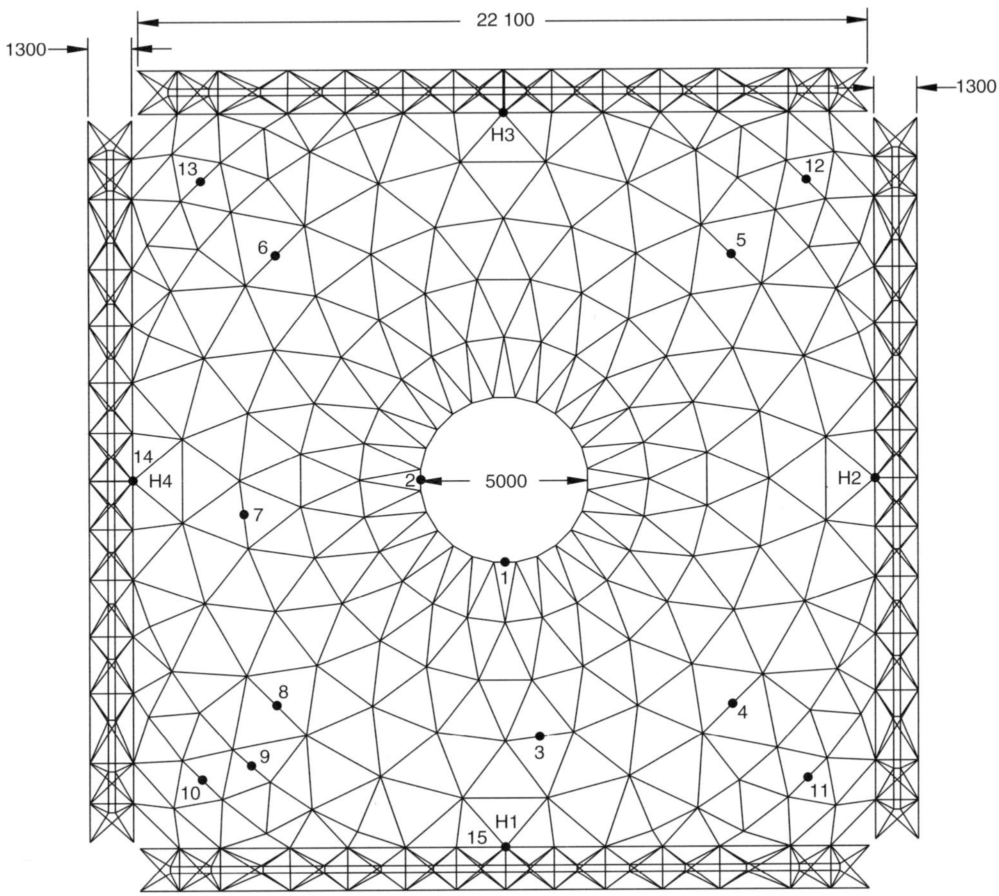


*Fig. 10.11. The measuring points of the revised design for the Pragati Maidan domes, New Delhi*

## 10.8.2. Proposed design modifications

In developing a revised design, it was considered prudent to aim at a snap-through load of 5 $kN/m^2$. Noting that the snap-through load is inversely proportional to the cube of the radius, the required radius is found as

$$\left(\frac{1.345}{5}\right)^{1/3} \times 22 = 14.18 \text{ m}$$

In consultation with the architect, the radius of the dome was eventually reduced from 22 m to 18.5 m, although this reduced the factor of safety against snap-through to 1.50. The alternative proposal of a double-layer dome was unacceptable to the architect. The other design modification proposed was to reduce the number of latitudinal rings from nine to four. Taking into account the views of the architect, six rings were provided in the revised design. The dimensions adopted in the revised design and a plan are shown in Figs 10.10 and 10.11. A typical member with a node connector is shown in Fig. 10.12.

### *10.8.3. Pilot test*

A pilot test was carried out on the prototype dome at the NCTL factory at Hyderabad. Some geometrical deviations in the region of the pendentives were detected by visual examination even before the test was commenced. These imperfections were attributable to the following factors:

- The assembly of the dome was carried out with frequent interruptions caused by heavy rain.
- The contractor assembled the dome proceeding from the top to the bottom in the following sequence:
  - the boundary arches were erected on the columns
  - using a central derrick, the dome was assembled top downwards adding one ring at a time
  - the pendentives were finally forced into position to rest on the arches. In this process, the arches also had to be forced to be seated on the columns. These adjustments inadvertently introduced initial geometrical imperfections.

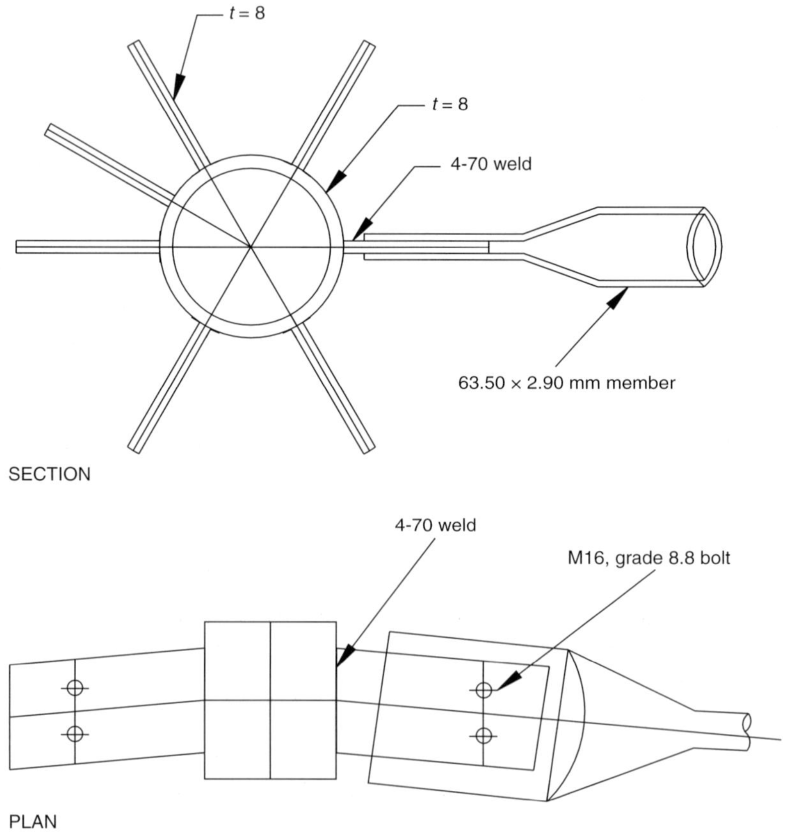


*Fig. 10.12. A typical member with a node connector (bolted connection) for the revised design of the Pragati Maidan domes, New Delhi*

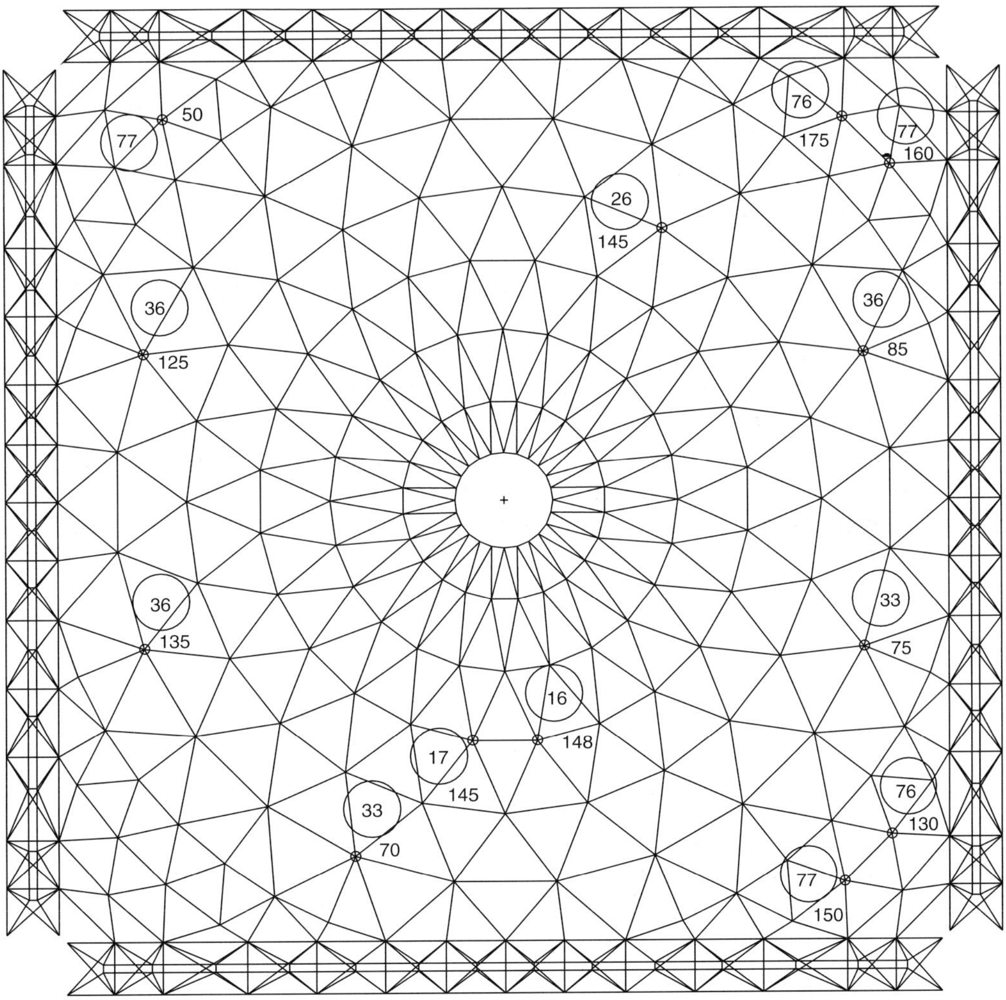


*Fig. 10.13. The vertical deflections measured at the observation stations during the pilot test for the Pragati Maidan domes (node numbers are enclosed in circles)*

As all the authorities concerned who had travelled to Hyderabad from New Delhi were already at the site, the test could not be postponed and had to be proceeded with. The positions of the observation stations where deflection measurements were taken are indicated in Fig. 10.13.

The dome was progressively loaded by stacking bricks on specially designed cradles hung from the nodes. Nearly 40 people were employed to carry the bricks and stack them on the cradle. A test on this scale is inconceivable in developed countries, because it would be prohibitively expensive. On reaching 60% of the design load, excessive deflections were noticed at some observation stations. These were most pronounced at nodes 76 and 77 in the region of the pendentives. Clearly, these were distress signals. The test was halted and the observed behaviour was referred to Dr Eekhout and his advice sought. The test was called off to be continued at a later date after carrying out any corrective measures that may be found necessary. The vertical deflections measured at the observation stations are shown in Fig. 10.13.

### *10.8.4. Modifications to the design and the erection sequence*

In the light of the observations made during the pilot test, the following design modifications and changes in the erection sequence were made:

- Because a single-layer pin-jointed dome is unduly sensitive to even small initial imperfections, it was decided to reinforce the connections with welds as indicated in Fig. 10.14. The change was inspired by the perceptive observation of Wright [10.2] that 'without bending strength and rigidity at the nodes or joints, snap-through buckling will occur under small loads; but with effective joints the

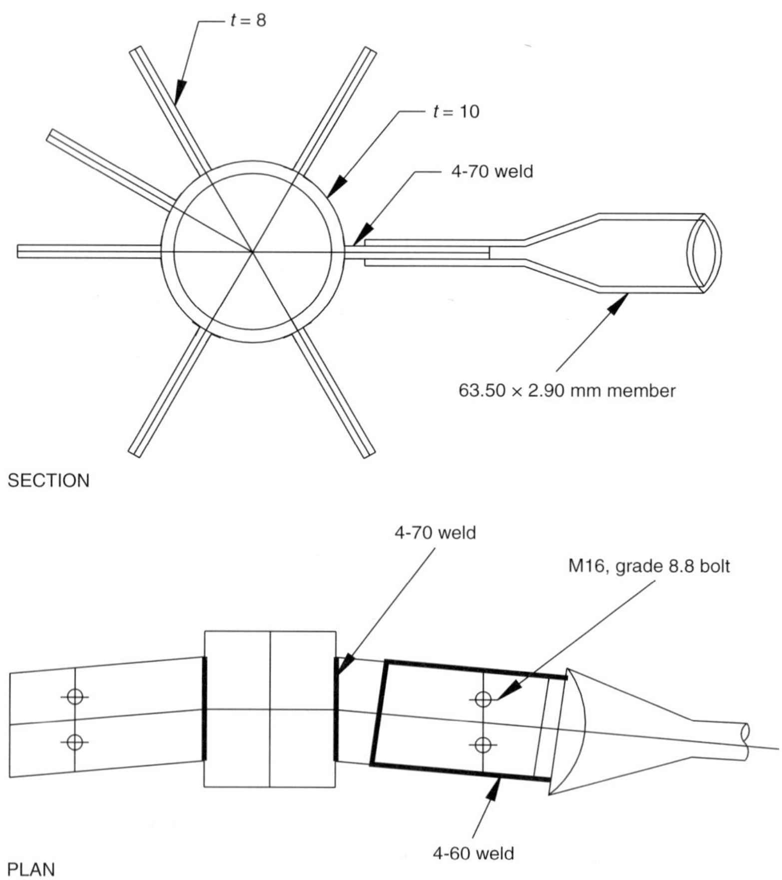


*Fig. 10.14. A typical member with a node connector (welded connection) for the revised design of the Pragati Maidan domes*

problem disappears'. The introduction of these welds at the joints made them behave in a moment-resistant fashion in the $M_1$ direction and in a pin-jointed fashion in the $M_2$ direction (Fig. 10.15).

- An additional eye ring, 2.4 m in diameter, was provided to improve rigidity. It was connected to the 5 m diameter ring by a triangulated assembly of the tubes (Fig. 10.16).
- The erection sequence was radically changed. The dome was erected from bottom to top instead of from top to bottom. The pendentives were first positioned on the arches. The domes with five rings, including the 5 m diameter eye, were assembled on the ground, thus permitting more accurate control over geometry. The co-ordinates of the erected dome were measured to ensure that they did not deviate from the computed co-ordinates.
- In some members, the two grade 8.8, M12 bolts were replaced by three grade 10.9 bolts.
- In some other members, the two grade 8.8, M12 bolts were replaced by two grade 10.9 bolts.
- In the dome of the pilot test, the two top chords of the boundary arches were connected by spacer rods. These were replaced by tubes to improve rigidity.

### *10.8.5. Final load test on the prototype dome*

#### *10.8.5.1. Domes and arches*

The prototype dome was an exact replica of the domes to be built at Pragati Maidan, New Delhi. The details of the supports of the arches over the columns were faithfully reproduced in the load test. It was not, however, possible to replicate the columns exactly, because some of them were 16 m high. To facilitate testing and taking deflection measurements, the column heights were scaled down to 1 m above the foundations.

#### *10.8.5.2. Loading method and sequence*

Loads were applied in a symmetrical sequence from the nodes through rods carrying steel cradles stacked with bricks. The design load of 170 kg/m$^2$ was applied in five equal increments. Once the full design load had been reached, it was sustained for 36 hours before unloading began in five equal decrements.

#### *10.8.5.3. Deflection observations*

Deflection observations were made at the stations indicated in Fig. 10.11. Observations were made between nodes in order to avoid interference with the loading rods located at the nodes. Vertical deflections were measured by means of specially fabricated devices that comprised vertical rods welded to the dome members. These devices telescoped into vertical tubes, carrying scales, fixed firmly to the ground. The measured deflections were cross-checked by means of simple water levels.

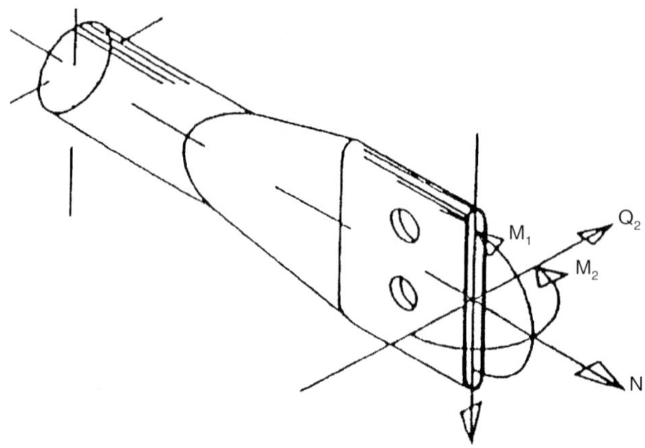


*Fig. 10.15. In the revised design of the Pragati Maidan domes the member connections were fixed with respect to M1 and pin jointed with respect to M2*

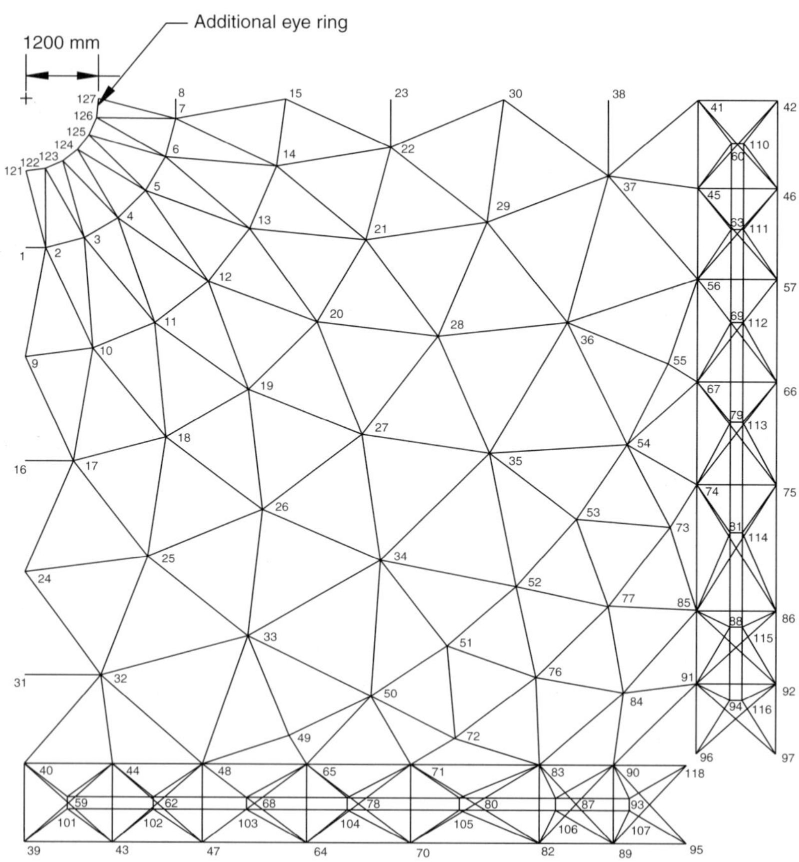


*Fig. 10.16. An additional eye ring of 2.4 m diameter improved the rigidity of the Pragati Maidan domes*

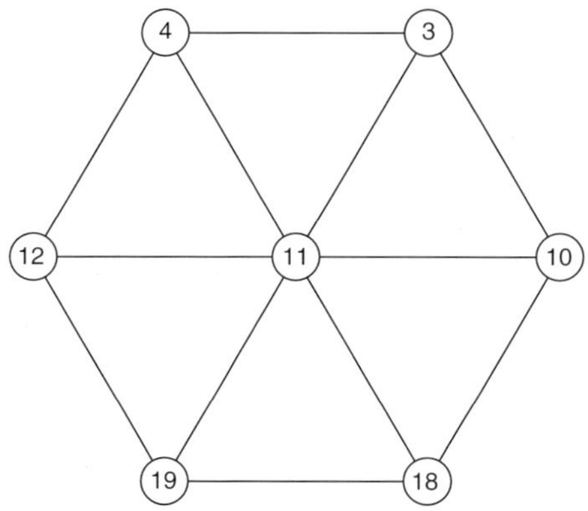


*Fig. 10.17. An element of the prototype dome*

### 10.8.5.4. Observations and conclusions

- No member or connector failed or showed any signs of distress.
- The maximum measured deflection of 17 mm was very much less than span/325 = 56 mm permitted by the Indian Code of Practice for General Construction in Steel (IS 800: 1984).
- The measured horizontal deflections were negligible.
- There was hardly any increase in the deflections under sustained loading.
- On unloading there was practically no residual deformation, indicating total elastic recovery.
- The dome was declared to have successfully passed the load test by meeting all the specified acceptance criteria.

## 10.8.6. Snap-through load computed by linear analysis using Wright's formula

An element of the dome bounded by nodes 4, 3, 10, 18, 19 and 12 and loaded at node 11 is shown in Fig. 10.17. The co-ordinates of node 11 are $x = 2144$, $y = -3714$, $z = 17996$. Although the length of all lines radiating from node 11 and connecting the boundary nodes are not exactly equal, they were assumed to be so to simplify the calculations. The co-ordinates of node 10 are $x = 1110$, $y = -4142$, $z = 17996$.

The length of member 11–10 is 1119 mm and the area of cross-section of the tubes $A$ is 820 mm$^2$. Hence the lower band snap-through load $q_{cr}$ is computed as

$$q_{cr} = \frac{820 \times 210 \times 1.119}{12 \times 18.5^3} = 2.54 \text{ kN/m}^2$$

and the working load is 1.70 kN/m$^2$. Hence the factor of safety against snap-through is 2.54/1.7 = 1.50.

*Note*: Before the dome was actually built, the design load was reduced to 90 kg/m$^2$ by a change of cladding. This ensured a higher safety factor against snap-through.

## 10.8.7. Non-linear analysis

### 10.8.7.1. Formulation

For a lattice dome structure, non-linear analysis is mandatory. The non-linear analysis used incorporates both geometric and material non-linearity. Adopting an updated Lagrangian frame of reference and an appropriate tensor definition for the stress and strain increments, the element tangent stiffness matrix is described using three contributions. These are: the conventional linear stiffness matrix; a geometric stiffness matrix, which introduces the effect of initial stress; and a deformation stiffness matrix, which accounts for the necessary coupling between the membrane and flexural and torsional actions within the element [10.16]. A lumped plasticity model is used to model material non-linearity. In this approach any plastic behaviour is deemed to be concentrated at the two extremities of an element. The stress resultants in the cross-section interact with each other to produce yielding of the section. This interaction is achieved using a yield surface in force space. The tangent stiffness matrix is augmented by a plastic reduction matrix, which is derived using the associated flow theory of plasticity [10.17].

An incremental iterative solution method is used with an arc length constraint. An additional constraint to control the force drift from the convex yield surface has to be imposed when considering material non-linearity. Rather than modifying the global stiffness matrix to reflect the extent of non-linearity, the load vector is modified by a residual force to maintain equilibrium. This can result in a substantial saving in computational time for large structures and, in addition, a more robust algorithm can be achieved.

This non-linear approach was used, as described below, to predict the structural response of the Pragati Maidan domes.

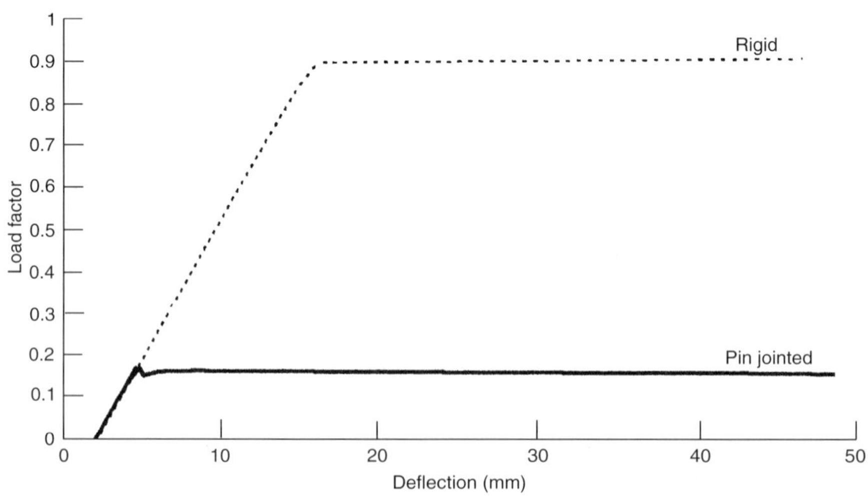


*Fig. 10.18. Load factor versus deflection curves for the rigid and pin-jointed domes*

*Table 10.1. The effect of using the rigid and pin-jointed models*

| | Station 1 | Station 3 | Station 4 | Station 11 |
|---|---|---|---|---|
| Test | 17.0 | 16.5 | 10.0 | 4.2 |
| Pin | 14.614 | 12.253 | 7.443 | 0.490 |
| Rigid | 14.237 | 12.711 | 7.395 | 0.476 |

### *10.8.7.2. Numerical results*

First, a non-linear analysis was undertaken of the dome with an eye of 5 m diameter used in the pilot test. A rigid model and a pin-jointed model were used in the analysis. The pin-jointed model showed that the dome buckled very soon after the incremental application of the load.

Buckling took place at a load factor of 0.18 (a load factor of 1.0 corresponds to a design load of 170 kg/m$^2$), followed by a rapid growth in displacement with a marginal change in load. The rigid-joint model predicted buckling at a load factor of 0.91, followed by a similar response to that of the pin-joint model. Figure 10.18 shows the load–deflection curves for both models at station 1.

During the pilot test, excessive deflections were observed at load factor of 0.6 and the test was halted. Given the fact that the joints will respond in a semi-rigid fashion, rather than as an idealized pin or rigid model and the presence of imperfections, the obtained results are in good agreement with the observations of the pilot test.

Next we analysed the modified dome with an additional eye ring of 2.4 m diameter. As before, both a rigid model and a pin-jointed model were used. Table 10.1 compares the computed deflections obtained with the two models at various locations on the dome, at a load factor of 1.0 (i.e. a design load of 170 kg/m$^2$), with measurements obtained during the physical test. There is reasonable agreement between the analysis and the test results, given the fact that imperfection was not modelled in the analysis (because no such measurements were made during the test). The effect of using a rigid or a pin-jointed model on the deflection is not that great. However, this effect is more pronounced when it comes to the stability of the dome model. The pin-jointed model predicted a snap-through buckling at a load factor of 1.43, while the rigid-jointed model predicted a value of 2.38. Using Wright's analytical method, snap-through will take place at $q = 2.55$ kN/m$^2$ using the pin-jointed idealization. This corresponds to a load factor of 1.5, compared with the value of 1.43 obtained using the numerical approach. The results of the non-linear analysis vindicate the intuitive decision taken to reinforce the joints by welding.

## 10.9. References

10.1. MEDWADOWSKi, S. J. Stability of shells and spatial structures. *Bulletin of the International Association for Shell and Spatial Structures*, 1979, **XX-3**(71), **XX1-1**(72).

10.2. WRIGHT, D. T. Membrane forces and buckling in reticulated shells. *Journal of Structural Division, Proceedings of the ASCE*, 1965, **ST1**(4227), 173–201.

10.3. Analysis, design and realization of space frames. Report of Working Group No. 8. *Bulletin of the International Association for Shell and Spatial Structures*, 1984, **XXV-1/XXV-2**(84/85), 62.

10.4. HANGAI, Y. and KAWAMATA, S. Non-linear analysis of space frames and snap-through buckling of reticulated structures. *Proceedings of the IASS Pacific Symposium on Tension Structures and Space Frames*, Tokyo and Kyoto, 1972.
10.5. MACLEISH, A. and BURGESS, I. W. A perturbation method for the non-linear analysis of reticulated shells. *Bulletin of the International Association for Shell and Spatial Structures*, 1979, **XX-1**(69), 25–33.
10.6. RIKS, E. An incremental approach to the solution of snapping and buckling problems. *International Journal of Solids Structures*, 1979, **15**, 529–551.
10.7. RIKS, E. Some computational aspects of the stability analysis of non-linear structures. *Computer Methods in Applied Mechanics and Engineering*, 1984, **47**, 219–259.
10.8. MALLET, R. H. and MARCAL, P. V. Finite element analysis of non-linear structures. *Journal of Structural Division ASCE*, 1968, **94**(ST9), 2081–2105.
10.9. RAJASEKARAN, S. and MURRAY, D. W. Incremental finite element matrices. *Journal of Structural Division ASCE*, 1973, **99**(ST12), 2423–2436.
10.10. RAJASEKARAN, S. Stability of reticulated structures revisited. *Technology, A Journal of the PSG College of Technology and Polytechnic*, 1998, 5–10.
10.11. HANGAI, Y. Numerical analysis in the vicinity of critical points by the generalized inverse. *Bulletin of the International Association for Shell and Spatial Structures*, 1995, **XXVIII-3**, 23–26.
10.12. RAMM, E. Strategies for tracing the non-linear response near limit points. *Nonlinear Finite Element Analysis in Structural Mechanics*, Springer-Verlag, Berlin, 1981, 63–69.
10.13. SAMARTIN, A. Numerical methods in non-linear analysis of shell structures. *Bulletin of the International Association for Shell and Spatial Structures*, **34**(112, No. 2), 81–102.
10.14. CHOONG, K. K. and HANGAI, Y. Review of methods of bifurcation analysis for geometrically non-linear structures. *Bulletin of the International Association for Shell and Spatial Structures*, 1993, **34**(2), 133–149.
10.15. KANI, M. I. and MCCONNEL, R. E. Single layer shallow lattice domes: analysis, general behaviour and collapse. *International Journal of Space Structures*, 1988, **3**(2), 64–73.
10.16. ALBERMANI, F. and KITIPORANCHAI, S. Non-linear analysis of thin-walled structures using least element/member. *Journal of Structural Engineering ASCE*, 1990, **116**(1), 215–234.
10.17. ALBERMANI, F. and KITIPORANCHAI, S. Elastoplastic large deformation analysis of thin-walled structures. *Engineering Structures*, 1990, **12**(1), 28–36.

# Index

Note: page references in *italics* refer to figures; those in **bold** refer to tables.